# Wind-Borne Debris Hazards

Subcommittee on Wind-Borne Debris

Edited by
**Nigel Berkeley Kaye, Ph.D.**

Sponsored by
Environmental Wind Engineering Committee of the
Wind Engineering Division of the
American Society of Civil Engineers

ASCE

**Published by the American Society of Civil Engineers**

**Library of Congress Cataloging-in-Publication Data**

Names: Kaye, Nigel Berkeley, editor. | American Society of Civil Engineers. Subcommittee on Wind-Borne Debris. | American Society of Civil Engineers. Environmental Wind Engineering Committee, sponsoring body.

Title: Wind-borne debris hazards / Subcommittee on Wind-Borne Debris ; edited by Nigel Berkeley Kaye, Ph.D. ; sponsored by Environmental Wind Engineering Committee of the Wind Engineering Division of the American Society of Civil Engineers.

Other titles: Windborne debris hazards

Description: Reston, Virginia : American Society of Civil Engineers, [2017] | Includes bibliographical references and index.

Identifiers: LCCN 2017048200 | ISBN 9780784414965 (soft cover : alk. paper) | ISBN 9780784481134 (PDF)

Subjects: LCSH: Buildings–Natural disaster effects. | Wind-borne debris–Risk assessment. | Wind damage.

Classification: LCC TH441 .W56 2017 | DDC 690/.92–dc23 LC record available at https://lccn.loc.gov/2017048200

Published by American Society of Civil Engineers
1801 Alexander Bell Drive
Reston, Virginia 20191-4382
www.asce.org/bookstore | ascelibrary.org



*Errata*: Errata, if any, can be found at https://doi.org/10.1061/9780784414965.


ISBN 978-0-7844-1496-5 (print)
ISBN 978-0-7844-8113-4 (PDF)
ISBN 978-0-7844-8116-5 (ePub)
Manufactured in the United States of America.

25 24 23 22 21 20 19 18 1 2 3 4 5

# Contents

# Acknowledgments


## Task Committee on Wind-Borne Debris

The Task Committee on Wind-Borne Debris was established by the Environmental Wind Engineering Committee of the ASCE Wind Engineering Division (WED). The goal of the committee was to produce a review of the current state of knowledge on issues related to risk from wind-borne debris from severe wind storms. The committee members worked with an international team of authors and reviewers to produce this state-of-the-art report on wind-borne debris hazards. The final peer reviewed document was approved by the Executive Committee of the WED on April 5, 2017.

**Task Committee Members:**
Nigel Berkeley Kaye, *Chair and Editor*
Gregory Alan Kopp, *Co-Editor*
Christopher William Letchford
Weichiang Pang
David Otway Prevatt
Douglas Alexander Smith

# Nomenclature

| | |
|---|---|
| $A$ | Debris cross-sectional area |
| $A_e$ | Critical condition coefficient |
| $A_p$ | Paver area |
| $\vec{a}$ | Acceleration vector |
| $C_a$ | Empirical asymptotic debris speed coefficient |
| $C_D$ | Drag coefficient |
| $C_f$ | Force coefficient |
| $C_L$ | Lift coefficient |
| $C_m$ | Moment coefficient |
| $C_p$ | Pressure coefficient |
| $C_{pB}$ | Pressure coefficient on base of paver |
| $C_{pT}$ | Pressure coefficient on top of paver |
| $C_z$ | Net uplift coefficient |
| $C_1$ | Asymptotic debris speed coefficient |
| $d$ | Debris diameter or smallest dimension of non-compact debris |
| $d_p$ | Particle diameter |
| $d_*$ | Dimensionless particle diameter |
| $DC_p$ | Uplift differential pressure coefficient |
| $F_D$ | Drag force |
| $F_L$ | Lift force |
| $F_R$ | Resisting force |
| $\vec{F}$ | Force vector |
| Fr | Froude number |
| $g$ | Gravitational acceleration |
| $h$ | Parapet height |
| $I$ | Mass moment of inertia |
| $K$ | Tachikawa number |
| $K_c$ | Critical Tachikawa number |
| $L$ | Longest debris dimension |
| $L_B$ | Building length |
| $l_f$ | Full-scale dimension |
| $l_l$ | Laboratory-scale dimension |
| $M$ | Moment |
| $m$ | Debris mass |
| $p$ | Pressure |
| Re | Reynolds number based on the skin friction velocity |
| $t$ | Time |

| | |
|---|---|
| $\bar{t}$ | Non-dimensional time |
| $U$ | Horizontal wind speed |
| $U_c$ | Velocity required to initiate flight |
| $U_{\mathrm{eave}}$ | Eave height wind speed |
| $U_f$ | Full-scale wind speed |
| $U_H$ | Mean wind speed at the mean roof height |
| $U_l$ | Laboratory-scale wind speed |
| $U_r$ | Reference velocity |
| $u_*$ | Skin friction velocity |
| $u_{*_c}$ | Critical skin friction velocity |
| $\hat{u}_H$ | 3 second gust speed at the mean roof height |
| $u_{\mathrm{rel}}$ | Horizontal component of relative wind speed |
| $V$ | Vertical wind speed |
| $V_I$ | Debris impact velocity |
| $V_R$ | Local 0.2 second gust speed at 10 m over flat terrain |
| $V_{\mathrm{rel}}$ | Relative wind speed |
| $v_{\mathrm{rel}}$ | Vertical component of relative wind speed |
| $W$ | Plate width |
| $x$ | Horizontal coordinate |
| $\vec{x}$ | Displacement vector |
| $\bar{x}$ | Non-dimensional horizontal coordinate |
| $z$ | Vertical coordinate |
| $\bar{z}$ | Non-dimensional vertical coordinate |
| $\alpha$ | Initial launch angle |
| $\beta$ | Angle between drag force and the horizontal coordinate |
| $\Delta$ | Square of the ratio of debris length to radius of gyration |
| $\theta$ | Angle of debris axis to the horizontal |
| $\xi$ | Angle of attack |
| $\rho_a$ | Density of air |
| $\rho_f$ | Fluid density |
| $\rho_p$ | Particle density |
| $\rho_{pf}$ | Full-scale debris density |
| $\rho_{pl}$ | Laboratory-scale debris density |
| $\rho_w$ | Density of water |
| $\tau_c$ | Critical surface shear stress |
| $\tau_0$ | Surface shear stress |
| $\upsilon$ | Kinematic viscosity |
| $\phi$ | Product of the lift coefficient and Tachikawa number |

# CHAPTER 1

# Introduction

Nigel Berkeley Kaye*

## 1.1 THE PROBLEM OF WIND-BORNE DEBRIS

Wind-borne debris in severe storms is a major hazard to people and structures. There is extensive literature on the damage to structures from wind-borne debris during hurricanes (e.g., Kareem and Bashor 2006) and tornadoes (e.g., Kuligowski et al. 2013). The mechanics of wind-borne debris is a complex problem involving motion initiation, flight through a heterogeneous unsteady wind field, impact, and resulting damage. In order to quantify the risk from wind-borne debris damage to a given structure one needs to understand potential sources of debris (e.g., tree limbs, roof aggregate, building components), the conditions under which they become wind-borne (e.g., limb breakoff, roof aggregate blow-off, building component failure), the flight trajectory (which depends on the release mechanism and the debris geometry), the energy and momentum at impact, and the ability of a structure to resist that impact. Much is known about each of these processes, although there is also considerable uncertainty about many of the key parameters. The prediction of the risk of wind-borne debris damage to a particular structure is hard to quantify.

There are standards for what portions of a building must be impact resistant for different wind climate regions. However, there is still much debate about the appropriate method for establishing these regions, the appropriate quantification of risk, and the resulting design response. This debate is clearly recognized in the commentary on enclosure classification, C26-10 of ASCE 7-10 (ASCE 2010), where the authors' state:

> The committee recognizes that there are vastly differing opinions, even within the standards committee, regarding the significance of these parameters that are not fully considered in developing standardized debris regions or referenced impact criteria.

Despite this ongoing uncertainty, significant research progress has been made in our understanding of all aspects of the wind-borne debris problem in recent years. The goal of this book is to provide a concise summary of our current state of knowledge on all aspects of the wind-borne debris problem along with detailed

*Glenn Department of Civil Engineering, Clemson University, Clemson, SC, USA.

reference lists for further reading. The book should prove valuable for engineering practitioners who are concerned with the risk of debris impact or debris mitigation. The book reviews the current state of knowledge and areas of uncertainty where risk is less well understood.

## 1.2 BOOK OUTLINE

The scale of the problem of wind-borne debris is laid out by David Prevatt and Dhruv Desai in Chapter 2. This chapter reviews the wind environment in which debris becomes wind-borne, the vulnerability of various structures to impact, responses to that risk, and the results of post-storm field surveys of debris damage to structures.

The next three chapters are structured to follow the path of a piece of wind-borne debris from motion initiation through flight to impact. In Chapter 3, Nigel Kaye and Arash Karimpour review the conditions under which debris begins to move. The chapter covers motion initiation for both naturally occurring debris, such as broken tree limbs, and engineered debris, such as failing roof shingles. Greg Kopp discusses the flight of wind-borne debris in Chapter 4. This includes the flight initial conditions, models for calculating debris trajectory, and the resulting impact conditions. The review discusses the three main debris geometries, namely compact, rod-like, and plate-like. The damage that results from debris impact on a structure is presented in Chapter 5 by David Henderson, Daniel Smith, Russell Carter, and Douglas Smith. They summarize models for impact damage and review current debris impact testing standards.

The final two chapters cover how the models and data from the earlier chapters are used to assess the risk of structural damage and how design codes have responded to this risk. Risk modeling for wind-borne debris is described by Michael Grayson and Weichiang Pang in Chapter 6. The chapter reviews the history of debris risk modeling and its evolution toward large-scale stochastic models that capture much of the physics of motion initiation, flight, and impact. Finally, Chris Letchford presents a detailed review of the current codes and standards that address the topic of wind-borne debris. Chapter 7 includes a discussion of the basis for writing the standards, the history and development of various codes and standards, and a detailed comparison of different testing codes and standards.

### *References*

ASCE. (2010). “Minimum design loads for buildings and other structures.” *ASCE/SEI 7-10*, Reston, VA.

Kareem, A., and Bashor, R. (2006). “Performance of glass/cladding of high-rise buildings in Hurricane Katrina.” *The Wind Engineer (Newsletter of AAWE)*, Dec., 1–5.

Kuligowski, E. D., Lombardo, F. T., Phan, L. T., Levitan, M. L., and Jorgensen, D. P. (2013). “Draft final report.” National Institute of Standards and Technology, Gaithersburg, MD.

# CHAPTER 2

# Overview and Field Data

David O. Prevatt*
Dhruv Desai†

## 2.1 INTRODUCTION

Substantial damage is caused to buildings and infrastructure by tropical cyclones, hurricanes, and typhoons, either by direct wind pressure–induced failure of building cladding and structural frames or by wind-borne debris and missiles carried along by fast-moving winds. Although studies over the past four decades provide a framework for evaluating the wind pressure effects on structures, understanding of and knowledge about wind-borne debris is somewhat less studied and more elusive. Wind-borne missile debris damage to building envelopes during hurricanes leads to major economic loss through damage to building contents, loss of lateral support, and collapse. Wind-borne debris is also an immediate hazard to building occupants.

Extensive building envelope damage is created by wind-borne debris after major wind events because building systems are often not designed to resist impact loads (Grayson et al. 2013). The extent of missile debris damage is variable and can be affected by the surrounding structures, or even by the amount of vegetation. In Florida, where the building code includes evaluation test protocols for impact resistance of building envelope systems, their performance in recent hurricanes has improved. Furthermore, the damage associated with specific missiles from an area can affect the extent of damage. Masters et al. (2010) observed that residential structures located in neighborhoods with predominantly tile roofing systems were twice as likely to experience window damage as homes located within neighborhoods with more asphalt shingle roofing.

Wind-borne debris is also prevalent in tornadoes, in which the damage is far more severe and extensive. The debris cloud produced around a tornado vortex is made of debris of all shapes and sizes swirling around the vortex and impacting buildings. It is common in tornadoes to observe motor vehicles thrown against

---

*Department of Civil and Coastal Engineering, University of Florida, Gainesville, FL, USA.
†Pfeil Design, Tampa, FL, USA.

*Fig. 2.1. Cars blown into buildings by a tornado*

buildings by the wind (see Fig. 2.1), in addition to rod-like (e.g., 2 × 4 lumber) and plate-like (e.g., plywood panels) wind-borne debris moving in the wind field.

As fast-moving missiles impact the building envelope, they breach the surfaces and create openings that allow wind-driven rain and air to enter the building. The resulting air pressurization and water damage to the building interior and its contents can be very costly. Air flowing into those openings increases the internal pressure and can cause failure of the roof structure, walls, or both (Sparks et al. 1994). Separating building damage from wind pressure versus wind-borne debris is a complex problem, and there is not an accepted means of distinguishing the economic consequences of each damage source. These challenges provide the rationale for studying debris production, flight, and damage, which is the purpose of this book.

The importance of wind-borne debris protection to building performance in hurricanes became quite evident following the devastating effects of Hurricane Andrew in 1992, which caused an estimated US$26 billion in damage. Enormous amounts of missile-like debris were created during this hurricane, causing substantial damage to building envelopes. Since then, engineers have had to address issues concerning building design to account for debris carried by the wind, manifested by changes to the building code design practice and test standards over the years.

The remainder of this chapter provides an overview of the damage resulting from wind-borne debris. The chapter starts by reviewing the wind environment in which debris can become wind-borne (Section 2.2) and the nature of loads developed by wind on the building envelope when breached by wind-borne debris (Section 2.3). Classification of debris on multiple bases is discussed in Section 2.4 which is followed by field data on the vulnerability of commercial and residential buildings to wind-borne debris (Section 2.5). Finally, damage mitigation methods (Section 2.6) are reviewed.

## 2.2 THE WIND HAZARD—HURRICANES, TORNADOES, AND STRAIGHT-LINE WINDS

There are different categories of winds based on their source of generation, nature, and behavior. Classifying them based on their nature, the wind hazards are primarily comprised of:

- Hurricanes, also called tropical cyclones or typhoons depending on where they occur, are non-frontal low-pressure systems of synoptic scale that develop over warm waters having organized convection and a maximum mean wind speed of 63 km/h (39 mi/h) or greater (in this book the term hurricane is typically used although the term tropical cyclone is used when discussing events, codes, and standards in Australia). As the event moves over warm ocean waters, it gathers heat and energy. Evaporation from the seawater increases its power to drive warm moist air toward the center of the storm that then spirals upward. As hurricanes strengthen they increase in size and intensity of wind speed. These events can continue to strengthen unless they come into contact with land or cooler water.

  In the North Atlantic, hurricanes form over warm ocean waters, usually starting as storms in the Caribbean or off the west coast of Africa. As the storms drift slowly westward, they gather heat and energy through contact with warm ocean waters. Hurricanes rotate in a counter-clockwise direction in the Northern Hemisphere around an "eye." The most violent part of a hurricane is near the center of the storm at the edges of its eye, called the eye wall. Hurricanes are defined as tropical storms having winds of at least 119 km/h (74 mi/h). Hurricanes strengthen until they come into contact with land or cooler water. When they come onto land, the heavy rain, strong winds, and heavy waves can damage buildings, trees, and cars. Fig. 2.2 shows a satellite image of a hurricane.

- Tornadoes are violently rotating columns of air extending from thunderstorms to the ground. The most violent tornadoes have rotating winds of 402 km/h (250 mi/h) or more. They are capable of causing extreme destruction, including uprooting trees and leveling well-made structures. They can turn normally harmless objects into deadly missile debris. Many tornadoes are just a few dozen meters wide and only briefly touch down, but highly destructive violent tornadoes may carve out paths over a mile wide and more than 80 km (50 mi) long. Although violent tornadoes comprise only 2% of all tornadoes, they are responsible for nearly 70% of tornado-related fatalities (source: National Disaster Education Coalition (1999); see Fig. 2.3).

- Straight-line winds such as microbursts or downbursts can also arise out of a thunderstorm. If these winds meet or exceed 93 km/h (58 mi/h), then the storm is classified as severe by the U.S. National Weather Service. These winds are produced by the downward momentum in the downdraft region of a

*Fig. 2.2. Satellite view of Hurricane Katrina, August 28, 2005*
Source: NASA (2005)

*Fig. 2.3. A tornado near Anadarko, OK; the lower part of this tornado is surrounded by a translucent dust cloud kicked up by the tornado's strong winds at the surface*
Source: National Oceanic and Atmospheric Administration (1999)

thunderstorm. Straight-line wind damage will push debris in the same direction the wind is blowing. Straight-line wind intensity can appear as powerful as a tornado. Because of this, some people in the general public will believe a tornado has occurred when in reality one did not occur. However, the debris field can be used to distinguish between tornadic and straight-line winds. Tornadoes tend to have relatively long and narrow damage paths, whereas straight-line winds tend to have wider, shorter damage paths.

## 2.3 WIND LOADS

The building envelope is defined as the components that enclose a structure. It consists of the cladding, roofing (roofing covering), exterior walls, glazing, door assemblies, windows, and skylights. Missile impact is the dominant cause of window glass breakage if there are wind-borne missiles, because winds tend to lift and sustain missiles at wind speeds considerably less than wind speeds required to cause critical pressure loads (Beason et al. 1984). When technical specifications and reports indicate that a particular building is designed for a particular wind speed, that design is normally based on the assumption of an intact building envelope; if the building is breached, then internal pressurization occurs and the wind load patterns change dramatically (HES Manual 1997).

When wind interacts with a building, a positive pressure is created on the windward wall, and negative pressure is created on the windward roof, leeward roof, and side walls (see Fig. 2.4). If the building envelope has a large opening on the windward wall, internal pressure becomes a significant fraction of the net loading on the other surfaces, with the internal pressure tending to push the walls outward and the roof upward.

The direction of internal pressure, inward or outward acting, depends on the location of major openings in relation to the direction of the wind. As shown in

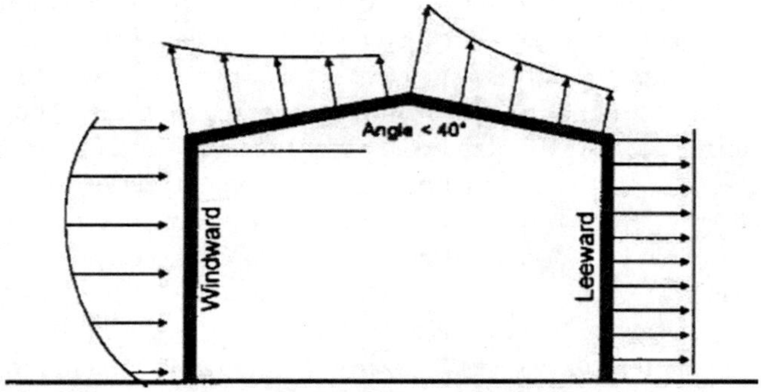


*Fig. 2.4. Typical wind pressure on low roof slope building*
Source: HES Student Manual (1997)

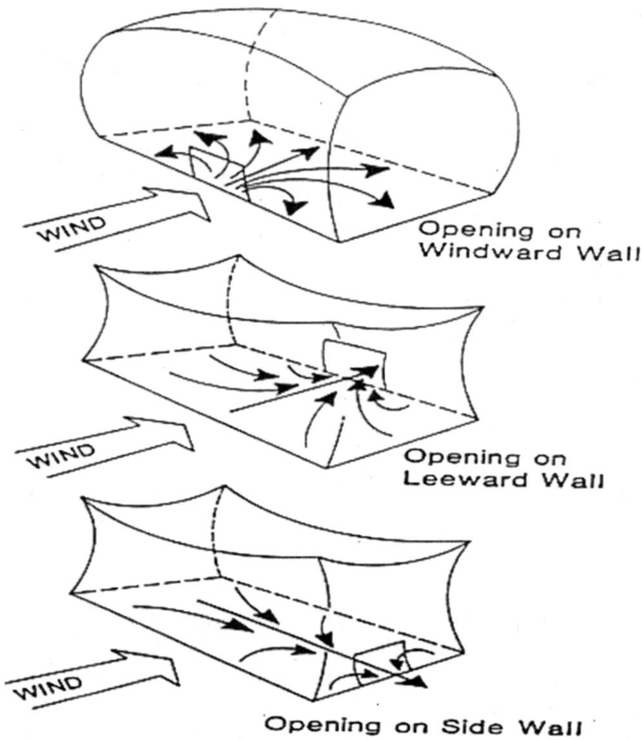


*Fig. 2.5. Internal pressurization and suction due to wall openings*
Source: HES Student Manual (1997)

Fig. 2.5, an opening that faces into the wind (windward wall) tends to pressurize the interior, thus amplifying loads in areas where the exterior loads are negative (outward acting) and decreasing loads in areas where the exterior loads are positive (inward acting). An opening in the walls parallel to the wind (leeward wall) or on side walls tends to cause negative internal pressure, which amplifies the net loads on the cladding in areas of positive exterior pressure and reduces them in areas of negative exterior pressure. Dominant openings (on the windward side) that are as small as 2% of the wall area will produce full internal pressurization (Kopp et al. 2008). Positive internal pressures can potentially double the surface pressures acting to lift the roof and push side and leeward walls outward. Fig. 2.6 shows the combination of external and internal pressure on the roof and the leeward wall face.

Hence, it is evident that a breach of the building envelope by wind-borne missile debris can produce several hazards including internal pressurization, exposure of occupants to wind and rain, damage to building contents, additional debris into the wind field, debris falling on the street, disrupted businesses, and a blemished appearance of the building envelope (Minor 1994).

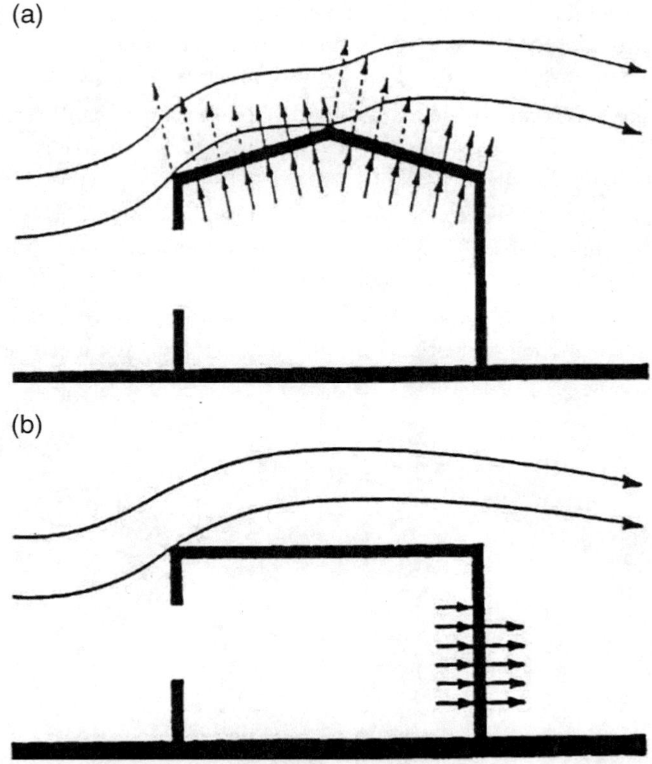


*Fig. 2.6. Increase in wind loads on walls and roofs due to internal pressurization: (a) combination of internal pressure with external suction on roof; (b) internal pressure combination with external pressure on leeward wall*
Source: HES Student Manual (1997)

## 2.4 DEBRIS CLASSIFICATIONS

Missile debris can be natural or manufactured objects composed of scattered fragments, typically of something wrecked or destroyed. Typical debris consists of roof tiles, roof gravel, tree limbs, signage, portions of damaged structures, framing lumber, roofing materials, and sheet metal. Debris can be classified as small or large (Minor 1994). Small missiles, such as roof gravel, can be carried to all elevations of a building facade at velocities large enough to break glass. Large missiles include a variety of items such as roof tiles, framing lumber, and tree limbs, and can impact the building envelope near ground level with sufficient force to penetrate walls and break windows and doors.

Wills et al. (2002) classified diverse debris geometrically into three types: "rod-like," "compact," and "plate-like." This classification identifies various debris types by their shapes and characteristic dimensions, which also affect their flight characteristics:

- Rod-like debris (linear debris) has one dimension significantly larger than the other two. Observed linear debris during wind storm events are light poles, street signs, metal rods, tree limbs, and lumber (see Fig. 2.7).
- Plate-like debris (planar debris) has one dimension significantly smaller than the other two. Planar debris includes roof sheathing, membrane commercial roofing, light siding material, and wood sheathing (see Fig. 2.8).
- Compact debris [three-dimensional (3D) debris] has similar dimensions in all three directions. Observed 3D debris includes vehicles, trees, furniture, trash containers, HVAC units, and gravel (see Fig. 2.9).

This rod, plate, and compact debris classification is the most commonly used classification for modeling debris flight. When considering damage from impact,

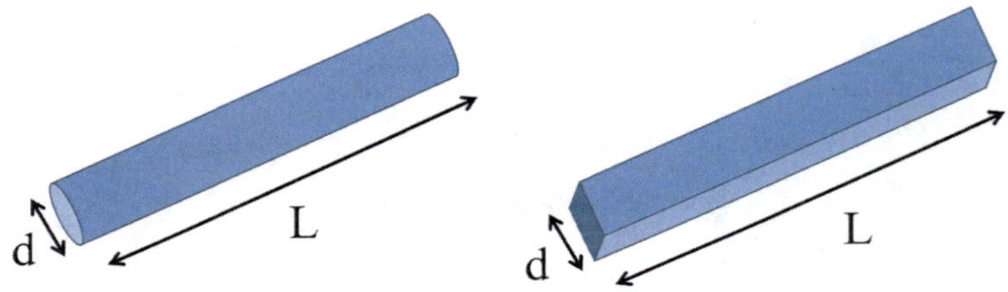


*Fig. 2.7. Rod objects showing dimensions ($L \gg d$)*

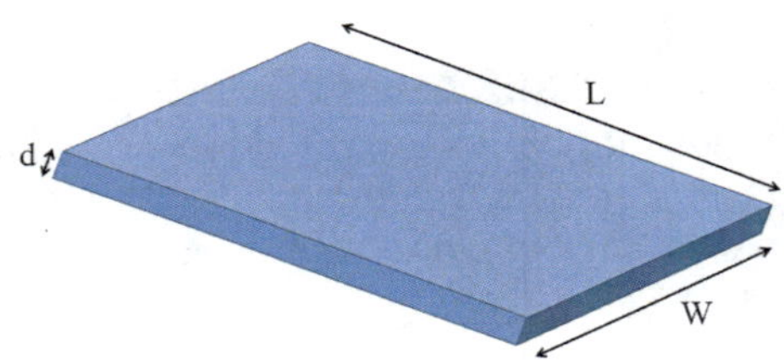


*Fig. 2.8. Plate-like (sheet) debris showing dimensions ($L,W \gg d$)*

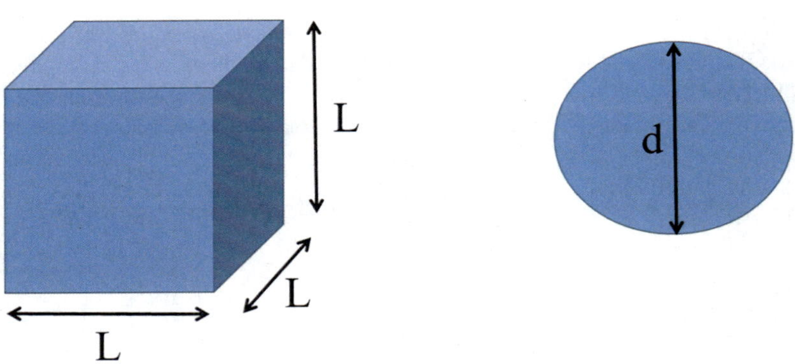


*Fig. 2.9. Compact objects showing dimensions*

*Table 2.1. Classification of Debris Based on Size and Weight*

| *Missile size* | *Debris* | *Associated damage* |
|---|---|---|
| Small (light) | Bricks, pieces of trees, roof aggregate roof surfacing, roof shingles, pieces of wood framing members | Broken windows, doors, light roof covering damage, damage to glazing |
| Medium (medium weight) | Long wood framing structures, plywood panels, steel decking, trash containers, furniture, appliances, HVAC units, steel pipes | Damage to walls, roof coverings, and roof structures |
| Large (heavy) | Structural columns, beams, joists, roof trusses, automobiles, trees, large tanks | Damage to roof and wall framing members and structural systems |

*Table 2.2. Classification of Debris Based on Location*

| *Location* | *Debris* |
|---|---|
| Urban and suburban areas/city | Roof gravel, window glass, membrane commercial roofing, metal commercial roofing, street signs, light poles, vehicles, roof sheathing |
| Rural area/country | Dirt, trees, broken pieces of trusses, wood sheathing, plywood, light siding material |

further classification of debris based on size, weight, location, and potential damage are also used. These are summarized in Tables 2.1 and 2.2.

## 2.5 VULNERABILITY OF BUILT INFRASTRUCTURE

The vulnerability of buildings depends significantly on the integrity of the building envelope. Whenever missile debris penetrates the building envelope, the building becomes more susceptible to further damage due to rain infiltration and increases in pressure inside the building. Hence, it is best to maintain the integrity of the envelope to reduce chances of significant damage. The following sections provide observations as to how commercial and residential buildings are vulnerable to extreme wind events and on the debris generated during these events.

### 2.5.1 Commercial Buildings

Commercial buildings have vulnerability to debris impacts because of the widespread use of glazing in commercial construction. This can yield large surface areas with brittle envelope material and can increase the probability of commercial buildings sustaining damage. The vulnerability of commercial buildings to wind-borne debris is discussed using a case study from Fort Worth, TX, which was hit by a tornado on March 28, 2000.

As shown in Fig. 2.10, the tornado passed through downtown Fort Worth near many commercial buildings ranging from four to 35 stories high. The most heavily damaged buildings were Cash America International, which was the first building in the tornado's path, followed by Educational Employees Credit Union, Mallick Tower, Lexington Place, Hunter Plaza, Tandy Center Outlet Mall Towers, and Bank One, with major cladding damage ending at UPR Tower. The debris accumulated because the cladding damage to each of these buildings increased the debris field as the tornado passed through the city. Buildings outside the main tornado path suffered less damage, mainly broken windows and marks left by wind-borne tarred-roof debris. Two building envelope failure mechanisms were observed: (1) inward displacement due to debris impact on the windward face and (2) outward displacement due to internal pressurization and possible internal debris impact.

The Cash America International building (see Fig. 2.11) had both kinds of failures. As the tornado passed, the southwest facade suffered the first impacts with breakage of glass on the southwest facade resulting from impact from internal debris, impact from external debris, and increased cladding load due to internal pressurization following window failure. The northeast facade was subjected to positive wind forces and debris impacts. Virtually every lite, both vision and spandrel, was fractured. Investigators believe debris impact caused most breakage on this facade.

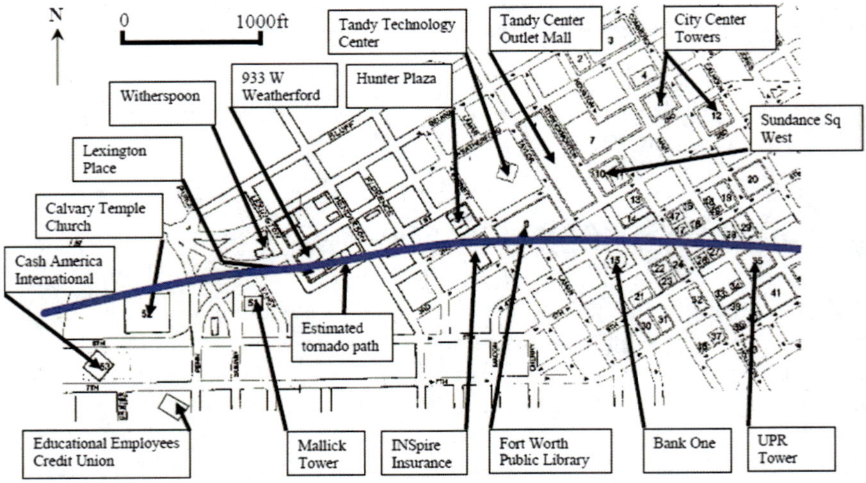


*Fig. 2.10. Map of downtown Fort Worth with traced tornado path*
Source: Letchford et al. (2000)

*Fig. 2.11. Cash America International building facade devastated by the tornado*
Source: Letchford et al. (2000)

The Educational Employees Credit Union, a 4-story building, suffered debris-impacted broken windows and loss of exterior insulation and finish system (EIFS) cladding. It is likely that a combination of suction and internal pressurization led to the EIFS cladding failures (extensive stripping of the cladding from the wall studs). Roof-mounted air conditioning units were displaced from mounts and the roof membrane was heavily damaged adjacent to corners where high suction pressures will have occurred.

Hunter Plaza, a 12-story U-shaped building, suffered window damage to all faces at all levels. The glazing was 3.175 mm (1/8 in.) single pane. The Inspire Insurance Building, which had two 6.35 mm (1/4 in.) panes separated by 12.7 mm (1/2 in.) as glazing and was right next to Hunter Plaza, provides an interesting case study. This building only suffered breakage of the outer panes, and hence the building envelope was not breached. This effective redundancy of the insulated glass units has major implications for recoverability of building function. Hence, it is evident from the tornado damage cited that commercial buildings have significant vulnerability and that redundant glazing can be helpful in preventing damage to the interiors of buildings. It is also important to note that storm shutters can also help protect the building envelope in such circumstances. Later sections describe preventive measures.

Recent damage to the St. John's Medical Complex in Joplin, MO (see Fig. 2.12) also highlights the effects of debris impact on commercial structures, where the entire fenestration of the building was destroyed in this 9-story reinforced concrete building. Although structural damage was minimal, the failed building envelope exposed the building interior to extensive damage, which caused failures of other systems. Eventually the hospital was condemned to be demolished, and a new hospital has since been constructed.

*Fig. 2.12. St. John's Regional Medical Center showing broken wind systems following the 2011 Joplin, MO, tornado*

### 2.5.2 Residential Buildings

Wind-borne debris from building roofs is a major cause of damage to architectural glazing systems (Beason et al. 1984). Researchers have found that a principal source of wind-borne debris is ridge shingles. The weakest portion of the residential building envelope is window glass, which is the most vulnerable to breakage in the event of debris impact. Breach of the envelope through window glass breakage would lead to internal pressurization, possible roof uplift, and extended damage due to rain and wind penetration.

Residential roofs are generally constructed using plywood sheathing over wood roof trusses and are covered with tarpaper and usually either extruded concrete tile, asphalt shingles, or clay. Based on a detailed study on damage to building envelopes in Dade County, Oliver and Hanson (1994) found few instances of wind-borne missile debris puncturing the roofs. The debris did not appear to be a significant cause of roof system failure, although most prevalent missiles in the residential region were clay and concrete tiles, which suggests that roof damage and expulsion of roof cover components was common.

The damage to roofs by debris impact is evident from the removal of roof coverings and dents in roofing materials but there have been few cases of roof structural system failure directly due to debris impact (on the roof). In fact, the impact of debris on a roof mainly creates more debris, which can potentially penetrate another part of the building envelope and cause roof uplift due to internal pressurization, which could lead to a roof structural system failure.

Exterior openings include double, single, and sliding glass entry doors, garage doors, windows, and skylights. A garage door failure can lead to significant damage (Fig. 2.13). It leads to internal pressure increases that contribute to roof failure and building envelope failures. Double entry doors, large sliding glass doors, windows, and skylights are most likely to be breached by debris. Doors and windows with shutters are observed to reduce damage by protecting the building envelope against debris penetration.

*Fig. 2.13. This garage door led to internal pressures that contributed to blowing out the roof and improperly connected wall corner joint of the garage*

In residential buildings, exterior walls are mostly siding (e.g. vinyl, wood, and fiber cement), concrete block or stucco, or EIFS insulation with light-weight, wood framing. Performance of masonry exterior walls after debris impact is generally better than that of wood frame structures with the siding, sheathing, insulation, and interior wall finish damaged by wind-borne debris impacts. Considering the wind-borne debris impact on building envelopes during Hurricane Andrew (1992) in Dade County, as shown in Table 2.3, the significance of debris impact on each of the various building envelope components is highlighted.

## 2.6 MITIGATION TECHNIQUES FOR WIND-BORNE MISSILE DEBRIS

Wills et al. (2002) developed a mathematical model for calculating damage potential (using a damage function consisting of the debris density, size, shape, and maximum speed attained) for flying debris in strong winds. Such research gives an understanding of how missile density and momentum influence damage to the building envelope. Based on past observations, together with damage-potential models, mitigation techniques have also been recommended and used to reduce damage from wind-borne debris. There are several ways to prevent or reduce damage. Storm shutters, to protect windows and doors, and laminated window glazing systems are two methods to protect windows from wind-borne debris impact. Storm shutters can adequately resist both impact loads and pressure loads, and hence are considered to be an effective protective system. They vary by characteristics and type of material used. Simple hurricane shutters are shown in Fig. 2.14.

*Table 2.3. The Degree of Damage Due to Missile Debris Impact on the Building Envelopes of Residential Buildings in Dade County (Based on 1992 Damage Surveys Following Hurricane Andrew)*

| *Envelope component* | *Type* | *Degree of damage observed* | |
|---|---|---|---|
| | | *Low* | *High* |
| Exterior wall | Masonry exterior wall | X | |
| | Wood frame structures with siding, sheathing, insulation, and interior wall finish | | X |
| Roof system | Plywood sheathing, underlayment with either extruded concrete tile, asphaltic composition shingles, or clay tile | X | |
| Envelope openings | Single-car garage doors | X | |
| | Two-car garage doors | | X |
| | Entry doors | | X |
| | Window system, including sliding glass doors, windows, and skylights | | X |
| | Storm shutters and boarded windows | X | |

Source: Oliver and Hanson (1994)

*Fig. 2.14. Simple hurricane shutters provide adequate window protection against missile debris*

## References

Beason, W. L., Meyers, G. E., and James, R. W. (1984). "Hurricane related window glass damage in Houston." *J. Struct. Eng.*, **110**(12), 2843–2857.

Department of Community Affairs/Division of Emergency Management and University of Florida/School of Building Construction (HES Student Manual). (1997). "State of Florida hurricane evacuation shelter selection guidelines student manual." ⟨http://www.floridadisaster.org/Response/engineers/HES.htm⟩ (Dec. 05, 2017).

Grayson, J. M., Pang, W., and Schiff, S. (2013). "Building envelope failure assessment framework for residential communities subjected to hurricanes." *Eng. Struct.*, **51**, 245–258.

Kopp, G., Oh, J., and Inculet, D. (2008). "Wind-Induced Internal Pressures in Houses." *J. Struct. Eng.*, **134**(7), 1129–1138.

Letchford, C. W., Norville, H. S., and Bilello, J. (2000). "Damage survey and assessment of Fort Worth tornado, 28 March 2000." Texas Tech Univ., Lubbock, TX.

Malaeb, D. A. (1980). "Simulation of tornado-generated missiles." M.S. thesis, Texas Tech Univ., Lubbock, TX.

Masters, F. J., Gurley, K. R., Shah, N., and Fernandez, G. (2010). "The vulnerability of residential window glass to lightweight windborne debris." *Eng. Struct.*, **32**(4), 911–921.

Minor, J. E. (1994). "Windborne debris and the building envelope." *J. Wind Eng. Ind. Aerodyn.*, **53**(1), 207–227.

NASA (National Aeronautics and Space Administration). (2005). "Hurricane Katrina (12L) approaching the gulf coast." ⟨http://visibleearth.nasa.gov/view.php?id=74693⟩ (May 2014).

National Disaster Education Coalition. (1999). "Talking about disaster: Guide for standard messages." ⟨http://www.disastercenter.com/guide/tornado.html⟩ (Dec. 05, 2017).

National Oceanic and Atmospheric Administration. (1999). "National Hurricane Center Outreach resources." ⟨http://www.nssl.noaa.gov/about/history/may3rd/⟩ (Dec. 05, 2017).

Oliver, C., and Hanson, C. (1994). "Failure of residential building envelopes as a result of hurricane Andrew in Dade County, Florida." *Hurricanes of 1992: Lessons learned and implications for the future*, R. A. Cook and M. Soltani, eds., ASCE, New York, 496–508.

Sparks, P. R., Schiff, S. D., and Reinhold, T. A. (1994). "Wind damage to envelopes of houses and consequent insurance losses." *J. Wind Eng. Ind. Aerodyn.*, **53**(1–2), 145–155.

Wills, J. A. B., Lee, B. E., and Wyatt, T. A. (2002). "A model of wind-borne debris damage." *J. Wind Eng. Ind. Aerodyn.*, **90**(4–5), 555–565.

# CHAPTER 3

# Debris Motion Initiation

Nigel Berkeley Kaye*
Arash Karimpour†

This chapter presents a review of the current understanding on how wind-borne debris becomes airborne. Research on the resulting flight, damage from impact, and resulting risk are discussed in later chapters. Wind-borne debris can be either naturally occurring (e.g., tree limbs, sand, and loose stones), or engineered (e.g., roof sheathing, shingles, and roof ballast such as pavers and aggregate). Section 3.1 discusses the basic theoretical modeling of motion initiation, regardless of source, the key non-dimensional parameter that controls motion initiation, and the scaling of laboratory studies to full scale. Section 3.2 reviews the literature on motion initiation of naturally occurring debris such as trees and loose stones. Roof aerodynamics is briefly reviewed in section 3.3 due to the influence of roof flows on a lot of manufactured debris. Section 3.4 describes the state of knowledge on motion initiation of debris from the built environment including roof pavers and aggregate, shingles, sheathing, and green roofs. Conclusions and directions for future research are presented in section 3.5.

## 3.1 MODELING AND SCALING BASICS

A simple theoretical model for motion initiation was presented by Wills et al. (2002). They theorized that an object in a wind field is exposed to wind loads and restraining forces. The wind loads can be lift

$$F_L = \frac{1}{2}\rho_a C_L A U^2 \tag{3.1}$$

and/or drag

$$F_D = \frac{1}{2}\rho_a C_D A U^2 \tag{3.2}$$

where $\rho_a$ is the air density; $C_L$ and $C_D$ are the lift and drag coefficients, respectively; $A$ is the object's frontal area; and $U$ is the wind speed.

---

*Glenn Department of Civil Engineering, Clemson University, Clemson, SC, USA.
†Louisiana Sea Grant, Louisiana State University, Baton Rouge, LA, USA.

For these equations there is a lot of information buried in the drag and lift coefficients. For example, the lift on a piece of roof gravel is controlled by the local wind speed, or in a different form, the local surface shear stress, $\tau_0$, often characterized in terms of a skin friction velocity

$$u_* = \sqrt{\frac{\tau_0}{\rho_a}} \tag{3.3}$$

However, the velocity typically used is some reference velocity such as the rooftop wind speed or the 10 m (32.8 ft) wind speed since these are used for wind design. For example, the value of $C_D$ measured in a wind tunnel includes information about the relationship between the upstream wind speed and the local surface shear stress on the roof. This, in turn, is controlled by the upstream terrain and the building geometry. Despite these complexities, the analysis of Wills et al. (2002) provides a useful conceptual framework and a mechanism for scaling laboratory results to full scale, provided information is also known about the restraining force, $F_R$.

For the case where debris is held in place by its weight, such as roof gravel or ballast pavers, then

$$F_R \sim mg \sim \rho_p L^3 g \tag{3.4}$$

where $m$ is the mass; $g$ is the gravitational acceleration; $\rho_p$ is the debris density; and $L$ is a characteristic length scale for (compact) debris. The exact nature of the restraint could be due to friction or restraint by objects downwind such as additional roof gravel. Regardless of the exact nature of the restraint mechanism (provided it is gravity based), the resulting condition for motion initiation is the same, namely, that the wind load exceeds the restraining force:

$$F_{D/L} > F_R \text{ or } \frac{1}{2}\rho_a C_F A U^2 \sim \frac{1}{2}\rho_a C_F L^2 U^2 > \rho_p L^3 g \tag{3.5}$$

where $C_F$ is a generic force coefficient. For given debris and topography, this result can be written in terms of the non-dimensional parameter

$$K = \frac{\rho_a U^2}{2\rho_p L g} \tag{3.6}$$

which is commonly known as the Tachikawa number (Holmes et al. 2006). Eq. (3.5) can then be written as $K > 1/F_F$. Therefore, provided geometric and dynamic similarities are satisfied, wind tunnel study results can be adjusted to full scale using

$$U_f = U_l \sqrt{\frac{\rho_{pf} L_f}{\rho_{pl} L_l}} \tag{3.7}$$

where the subscript $f$ indicates full-scale conditions and the subscript $l$ indicates laboratory-scale conditions. A detailed discussion of the conditions required for full dynamic and geometric similarity is presented on a case by case basis in the later sections.

For cases where the debris is not restrained by its weight, for example, shingles and tree branches, the restraining force is mechanical. Therefore, the Tachikawa scaling is not appropriate for initiation of motion, and a detailed model of the mechanical restraint is required. This could include quantifying the pull-out resistance of the nails holding down sheathing or shingles, or the shear and bending strength of a tree branch. Typically, this means that wind tunnel studies cannot be directly adjusted to full scale. Instead, wind tunnel studies are used to measure the aerodynamic forces, which are then combined with full-scale data on the restraining forces. These conditions are then discussed on a case by case basis.

## 3.2 NATURALLY OCCURRING DEBRIS SOURCES

Naturally occurring debris, as opposed to debris that originates from engineered structures, largely consists of plant matter and loose sand and stones. This section focuses specifically on trees rather than on all plants, since tree debris has much greater potential for damage. The section also reviews research on the motion initiation of sand and small aggregate since this can cause damage to structures. It also provides the background needed for examining roof gravel blow-off.

### 3.2.1 Trees

The failure of trees under wind loads is a complex problem. Trees come in a wide variety of species, sizes, shapes, and ages. The material properties of trees are also highly variable. The wind load on trees can vary with season due to the loss of leaves. Trees also adapt to their surroundings. For example, James et al. (2006) showed that trees grow additional wood support in areas of larger gravitational or wind load. Accordingly, a change in the prevailing wind flow around a tree, such as through the removal of surrounding trees or the construction of a building that alters the wind field, has the potential to make the tree unstable. Trees are also flexible, and thus aero-elastic interactions can occur. Given all this complexity, it is a difficult task to quantify both the wind loads and the restraining forces.

There has been research into various failure mechanisms including uprooting and trunk snapping (Putz et al. 1983) and branch breakoff, which is characterized as a defense mechanism for trees (James et al. 2006). However, given the vast variety of geometries and material properties, there is little in the way of generally applicable predictive failure data. Furthermore, most of the research in this area is focused on the forestry industry (Peltola and Kellomaki 1993; Putz et al. 1983) rather than on more isolated trees that are typical of developed areas.

The wind and gravitational loads that lead to tree failures have been addressed in various theoretical biomechanical models for tree behavior. Shigo

(1985) developed a model for the attachment of branches to tree trunks that gained wide acceptance. However, more recently, Slater and Harbison (2010) argued that Shigo's model is flawed and proposed an alternate attachment model. A simple damped beam model for branch behavior and failure for urban trees was proposed by Baker and Bell (1992). The model compares well with field tests. However, as with the models of Shigo (1985) and Slater and Harbison (2010), no systematic method for calculating the model parameters for a given tree are provided. The models are, thus, of limited practical value in predicting branch failure.

The wind load on a tree is also a complex nonlinear problem. The area of a tree normal to the wind flow changes with wind speed due to the flexing of the tree and the bending of leaves to align themselves to the wind direction. As such, the drag force on the tree does not vary with the square of the wind speed. A series of tests in which trees were dragged along a water channel indicated that, for low speeds, the trees did not flex. However, above a critical wind speed the trees adjusted to the wind speed and the drag force varied linearly with the wind speed (Wilson et al. 2010). However, other tests on tree drag exhibit a range of velocity exponents ranging between 1 and 2 (Kane and Smiley 2006).

The one area where there is a clear relationship between tree damage and wind speed is in post-disaster assessments of storm strength, such as the Fujita scale (Cullen 2002). However, this approach has significant limitations when used to predict debris formation. For example, the connection between wind speed and tree damage is largely empirical and does not address the type, size, or age of the tree. Furthermore, the wind speed bands are quite large, ranging in width from 63 to 92 km/h (39 to 57 mi/h). An attempt to improve the theoretical underpinnings of the connection between wind speed and tree damage has been made by Beck and Dotsek (2010) using the HWIND model (Peltola and Kellomaki 1993). The model was run using simulated tornado wind speeds to assess the damage to forests by two tornados in Spain and the Czech Republic. However, this study is focused on forests, rather than trees in urban areas.

In summary, while there has been much research into the failure of trees, we do not have predictive tools that can quantify the conditions under which a tree, or tree component such as a branch, will break off and become airborne.

### 3.2.2 Sand and Loose Stones

The motion initiation of sand and loose stone has been an area of research interest since the early twentieth century. Bagnold (1937, 1941) examined the conditions under which sand would become airborne using a series of wind tunnel tests on sand. The driving force for motion initiation was the wind-induced surface shear stress, $\tau_0$, and the buoyancy force due to the surrounding air. The restraining force is the particle weight. The resulting force balance scaling is given by

$$\tau_0 \pi d_p^2 + \frac{4}{3}\pi\rho_f g d_p^3 \sim \frac{4}{3}\pi\rho_p g d_p^3 \tag{3.8}$$

where $d_p$ is the particle diameter; $\rho_f$ is the fluid density (in this case, air); and $\rho_p$ is the particle density. This can be rewritten in terms of the critical skin friction velocity ($u_{*_c}$) as

$$u_{*_c} = A_e \sqrt{\frac{\rho_p - \rho_a}{\rho_a} g d_p} \tag{3.9}$$

where $A_e$ is an empirical constant found to be approximately $A_e = 0.1$ for particles larger than 0.2 mm (0.0079 in.). This can be rewritten as a densimetric Froude number (Fr) as follows:

$$\mathrm{Fr} = \frac{1}{2} \times \frac{\rho_a u_*^2}{(\rho_p - \rho_a) d_p g} = \left(\frac{A_e^2}{2}\right) \tag{3.10}$$

For solid particles in air the air density is significantly smaller than the particle density such that the density difference $(\rho_p - \rho_a) \approx \rho_p$.

A more recent study by Iversen and White (1982) showed that the critical condition for motion initiation (in other words, the value of $A_e$) is actually a function of the particle Reynolds number

$$A_e = A_e(\mathrm{Re}_*) \tag{3.11}$$

where

$$\mathrm{Re}_* = \frac{u_* d_p}{\upsilon} \tag{3.12}$$

and $\upsilon$ is the kinematic viscosity of the fluid (in this case, air).

Actually, the result of Iversen and White (1982) was already largely established from sediment transport studies (Shields 1936a, b, c) even before Bagnold's work was published. Although there are significant differences between sediment motion initiation in streams and sand removal due to wind, they do both involve particles being transported by a turbulent boundary layer, and have their motion initiated by the boundary layer surface shear stress.

Shields (1936a, b, c) investigated the relationship between the critical surface shear stress required to initiate sediment motion and the particle Reynolds number. Shields' results are presented in non-dimensional form as

$$\frac{\tau_c}{(\rho_p - \rho_w) g d_p} = \frac{\rho_w u_{*_c}^2}{(\rho_p - \rho_w) g d_p} = f\left(\frac{u_{*_c} d_p}{\upsilon}\right) \tag{3.13}$$

where $\rho_w$ is the density of water; $\tau_c$ is the critical surface shear stress for motion initiation; and

$$u_{*_c} = \sqrt{\tau_c / \rho_w} \tag{3.14}$$

is the critical skin friction velocity. This functional relationship is typically represented graphically on a Shields diagram as shown in Fig. 3.1.

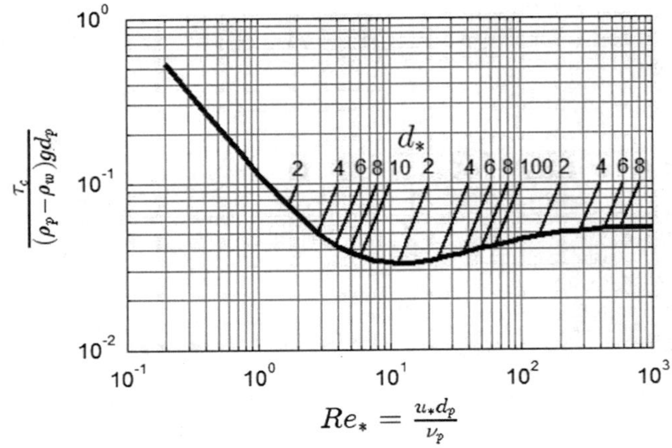


*Fig. 3.1. Shields' diagram*
Source: Guo (2002)

One problem with Eq. (3.13) and the Shields diagram representation of the critical condition for motion initiation is that the critical condition ($u_{*_c} = \sqrt{\tau_c/\rho_w}$) appears on both sides of the equation (both axes of the plot). The formulation is implicit and harder to use. To overcome this problem the dimensionless particle diameter can be used,

$$d_* = d_p \left( \sqrt[3]{\frac{(\rho_p - \rho_w)g}{\rho_w \nu^2}} \right) = \sqrt[3]{\frac{\mathrm{Re}^2}{\mathrm{Fr}}} \tag{3.15}$$

which is independent of the skin friction velocity and can be related directly to the critical shear stress (see Fig. 3.1). As such, the Shields diagram can be used to predict the initiation of motion of sand and small aggregate when forced by a fully developed turbulent boundary layer. However, the formulation in Eq. (3.13) requires knowledge of the skin friction velocity rather than a mean reference wind speed, $u_r$. The skin friction velocity can be related to a reference wind speed using a drag coefficient

$$C_D = \frac{u_*^2}{u_r^2} \tag{3.16}$$

although the drag coefficient $C_D$ can be hard to establish. Furthermore, the formulation only applies to flat terrain, and should not be applied to more complex geometry such as loose-laid gravel on a built-up roof. Note that in all the equations used in this section, $\rho_w$ and $\rho_a$ can be taken as $\rho_f$, the density of the fluid.

## 3.3 REVIEW OF ROOF AERODYNAMICS

The built environment provides numerous sources of debris from building materials that are detached from structures. In this case, the motion initiation is largely controlled by the aerodynamics of the air flow over and around the building. This section presents a brief review of building aerodynamics including air flow and pressure distributions, the role of parapets in mitigating extreme suction pressures, and potential scaling issues associated with laboratory-scale measurements of both pressure coefficients and debris motion initiation.

### 3.3.1 Flow Separation and Vortex Formation

Air flow around buildings is complex, three-dimensional, and turbulent. However, there are some simple underlying principles and phenomena that, broadly speaking, explain much of the wind-structure interaction.

As the wind flows over and around a building the flow path is deflected by the structure. For sharp-edged cuboid structures, or structures with low-slope roofs, this deflection leads to flow separation at the rooftop and side walls [see Fig. 3.2(a)]. This flow separation produces curved streamlines over the rooftop (and around the building sides). The air must, therefore, accelerate normal to the streamlines down toward the building roof (and side walls). This, in turn, induces low pressure on the roof surface since a pressure gradient is required to produce this acceleration. The resulting suction pressure on the roof will encompass the entire separation bubble formed at the building's leading edge. The suction pressure is parameterized in terms of a pressure coefficient. The flow separation and suction pressure is also observed for low-slope roofs, although for higher roof slopes there can be positive pressure on the windward side of the roof and suction pressure on the downwind side. For all cases the suction is particularly high near the leading edge. See ASCE 7-10 (ASCE 2010) for specific values of design pressure coefficients for a range of low-rise building geometries.

The exact flow paths taken over a building depend on the building geometry, the up-wind flow properties (turbulence intensities, turbulent integral length scales, and velocity profile), and the wind direction. One flow geometry of particular interest is the so-called cornering or quartering flow that occurs when the wind angle is not normal to the building walls. In such cases, conical vortices can form that are almost parallel to the building walls [see Fig. 3.2(b)]. These cornering or "delta-wing" vortices can produce very high suction pressures immediately below them with instantaneous point pressure coefficients as high as 6.5 (Stathopolous and Baskaran 1987). The consensus in the literature is that the suction pressures are greatest for flat and very low-slope roofs.

### 3.3.2 Role of Parapets

Parapets are often used to mitigate the suction forces on roofs. Stathopolous et al. (1999) showed that, in general, the presence of a circumferential parapet reduced the mean suction on a flat roof. However, if the parapet is not high enough it can

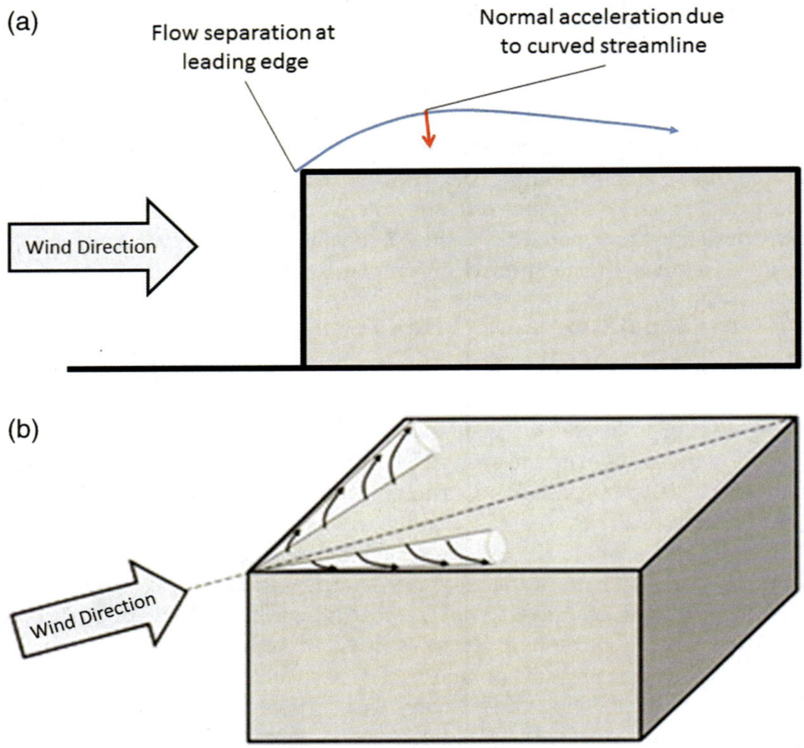


*Fig. 3.2. Schematic diagrams of (a) the flow separation over a flat roofed building showing the resulting curved streamline and path normal acceleration that induces the roof suction pressure, and (b) the steady conical delta wing vortex that forms from quartering flows*

induce larger suction pressures. Stathopolous et al. (1999) showed that, for a flat roof, increased suction is observed when the parapet height $h$ is less than 2% of the building length, $L_B$ (that is, $h/L_B < 0.02$). The increase in suction with small parapets was observed at both laboratory and full scale, indicating that it is not an artifact of small-scale wind tunnel testing. The increase in suction measured for low parapet heights was addressed by Kopp et al. (2005), who showed that the parapet raises and slightly deflects the corner vortices above the roof, allowing them to be more intense. This results in higher suctions for lower parapets. However, for higher parapets, the vortex is lifted far enough above the roof such that the suction is less than for the no parapet case. Kopp et al. (2005) also showed that parapets can lead to a positive (downward) pressure on the downwind end of the building due to stagnation of the reattached flow.

Isolated parapets, that is, parapets that do not completely enclose the circumference of the roof, produce flow separation regions at the vertical parapet edges (Kopp et al. 2005). This can increase uplift and move the location of peak suction. This is particularly true for high parapets. Porous parapets and porous corner screens and other edge fairings (Banks 2000) will inhibit the formation of vortices (Cochran and English 1997), and thus considerably reduce peak roof pressures.

Parapets have also been tested on sloped roofs. For example, Suaris and Irwin (2010) examined the role of parapets in reducing suction loads on a gable roof house. Their study examined corner, perimeter, and perforated parapets and found that they all reduced corner suction loads. Interestingly, the parapets considered were relatively short ($h/L = 0.01$), and as such would have resulted in higher suction pressures for flat roofs (see Stathopolous et al. 1999). This may be due to the slope of the roof inhibiting the growth of the conical vortices.

### 3.3.3 Scaling Effects in Experimental Studies

Much of the data on pressure distributions and debris motion initiation has been collected from small- to medium-scale wind tunnel experiments. It is, therefore, important to discuss their relevance at full scale. As discussed above, motion initiation occurs when a lift force or moment overcomes the debris restraint. In the event of loose-laid debris such as gravel, the balance between the initiating (wind-driven) and restraining (weight) forces is parameterized in terms of a densimetric Froude number (Fr) (Karimpour and Kaye 2012). For restrained objects such as shingles, the uplift pressure is parameterized using a pressure coefficient ($C_p$), which is then used to quantify the full-scale uplift force on the object. In both cases, the classical assumption of Reynolds number independence is used. However, there is increasing evidence that, even for flow around sharp-edged objects, some Reynolds number effects could influence motion initiation predictions that are based on reduced scale wind tunnel studies. Reynolds number effects on pressure coefficient measurements are discussed in this section. Reynolds number effects on studies of loose-laid debris are discussed in more detail in the roof gravel and roof paver sections below.

Richards et al. (2001) presented results from a full-scale test of the pressure distribution over a 6 m (19.7 ft) cube. The magnitude of the measured mean pressure coefficient at the center of the roof for a wind normal to the cube face increased with increasing Reynolds numbers. The results were compared to a range of small-scale wind tunnel tests (from the Wintechnologische Gesellschaft comparative testing program, Hölscher and Niemann 1998), which showed that the mean $C_p$ at small scale was roughly $C_p = -0.6$ at the roof center compared to roughly $C_p = -0.8$ for the full-scale tests. The level of disagreement between small and full scale increases in regions where there is flow separation and reattachment. More recently, Lim et al. (2007) presented a detailed analysis of the flow conditions required to accurately model the pressure distribution on a bluff body at small scale. They highlight the need to accurately represent the energy spectra when measuring pressure fluctuations on a bluff body. This results in an inherent Reynolds number effect since the range of scales in the energy spectrum will increase with increasing Re. Therefore, a correction for the use of small Reynolds numbers is required for measurements of fluctuating, but not mean, pressure coefficients. The Reynolds number effect is larger in regions with strongly concentrated or relatively steady vortex regions such as near the conical vortex that forms during quartering flows. In this case, the vortices are largely independent of the upstream flow turbulence characteristics but are dominated by the building geometry. Reynolds number effects on the mean and fluctuating pressure coefficient are clear near the vortex core. Unfortunately, no methodology for undertaking a Reynolds number effect correction was presented by Lim et al. (2007). Part of the reason for this is that there are larger challenges and uncertainties in wind tunnel testing of low-rise buildings that have not yet been fully resolved, as discussed by Tieleman (2003). In particular, there are concerns regarding missing low-frequency turbulent energy in wind tunnel spectra for low-rise buildings. There are currently a number of researchers working on theoretical corrections to overcome this problem (see Mooneghi et al. 2015; Banks et al. 2015).

## 3.4 WIND-BORNE DEBRIS FROM BUILDINGS

Wind-borne debris from buildings can be broadly categorized as either loose laid (e.g., roof pavers, green roof modules, and gravel ballast) or fixed (e.g., sheathing, metal roofs, and shingles). The motion initiation of each of these materials is reviewed in turn.

### 3.4.1 Roof Pavers

Roof pavers are placed on flat rooftops to provide ballast against wind suction loads, to provide physical protection for the roof membranes from wind shear and rainfall, and to prevent ultraviolet degradation of the roof sealant. They are usually placed on the roof and held in place by their weight. Clips can be used to connect sets of pavers to increase the weight of the objects that must be dislodged for the ballast to fail (which increases the effective area of the object if the clips are strong enough to withstand load transfers at the initiation of motion).

Early studies of roof paver blow-off used small-scale wind tunnel tests to establish the full-scale equivalent wind speed required to lift up a paver. Bienkiewicz and Meroney (1988) investigated the role of parapet height, paver configuration, and clips on the critical wind speed for failure. Failure quantification was based on visual observation of uplift from multiple runs under the same conditions. Bienkiewicz and Meroney (1988) established that a staggered grid of clipped pavers resulted in the highest wind speeds required for failure. They also showed that, for small parapet heights, the critical wind speed for blow-off was lower. This is qualitatively in line with the parapet suction measurements of Stathopolous et al. (1999) and Kopp et al. (2005) described above. However, the parapet height required to achieve the same blow-off speed as the no-parapet case was substantially higher than the parapet height required to produce a net reduction in suction as reported by Stathopolous et al. (1999). This may be in part due to the different length-to-building-height ratio tested by Bienkiewicz and Meroney (1988).

A physical explanation for the failure mechanism was investigated by Bienkiewicz and Sun (1992), who measured the pressure distribution above and below a set of model pavers in a wind tunnel. Failure occurred when the pressure difference multiplied by the paver area was greater than the paver weight. However, the relationship between the top and underside pressure distributions is complex and controlled by the rate of pressure equalization due to flow in the gaps between pavers and in the cavity between the paver and the roof surface. The pressures on the top and bottom of the paver are parameterized in terms of pressure coefficients $C_{pT}$ and $C_{pB}$, respectively. The net uplift is given by

$$F = \frac{1}{2}\rho_a U^2 (C_{pT} - C_{pB}) A_p \tag{3.17}$$

where $U$ is the reference wind speed; $\rho_a$ is the density of air; and $A_p$ is the paver area. Failure will occur when the uplift force exceeds the weight of the paver, that is, when

$$\frac{1}{2}\rho_a U^2 (C_{pT} - C_{pB}) A_p > mg \tag{3.18}$$

or

$$\mathrm{Fr} = \frac{\rho_a U^2 (C_{pT} - C_{pB}) A_p}{2mg} > 1 \tag{3.19}$$

where Fr is the paver densimetric Froude number.

The pressure equalization is more effective when the gap between the pavers is larger since there is less resistance to flow. This leads to a higher correlation between the top and underside pressures and improved performance (higher critical wind speed). On the other hand, as the gap between the bottom of the paver and the roof surface becomes larger, more air needs to be removed to equalize the pressures, which causes less-extensive equalization, a lower correlation in pressures across the paver, and lower performance.

Although the pressure equalization was relatively fast, high-frequency pressure fluctuations were highly attenuated across the paver. The pressure correlation across the pavers also increased with increasing parapet height although very small parapet heights were not investigated. An analytical model for the pressure equalization using Darcy's law under steady flow was presented in Sun (1993). A more systematic investigation of the role of paver spacing and the gap width under the paver on pressure equalization was conducted by Bienkiewicz and Sun (1997), who showed that as the ratio of the inter-paver spacing to the under-paver gap increases, the magnitude of the under-paver $C_p$ increases, the smaller is the pressure difference across the paver, and the better is the performance. Bienkiewicz and Sun (1997) also found that narrower pavers perform worse.

Full-scale testing was conducted on roof pavers by Aly et al. (2012). The study was conducted on an 8 ft square plan building for five different wind angles. The study measured the blow-off wind speed for full-scale pavers, although the gap under the pavers was larger than one might expect for normal installations. This would lead to reduced pressure equalization (Bienkiewicz and Sun 1992, 1997), which means that their measured wind speeds for blow-off may be lower than would be observed in the field. These authors did introduce a new failure mechanism. Rather than assuming failure occurs when the net uplift is greater than the weight of the paver, they suggested that motion is initiated when the moment about the downwind end of the paver, generated by the uneven suction pressure distribution over the paver, exceeds the moment about that point due to the paver weight. Again, they showed that clipping pavers together increases their resistance to wind-driven motion initiation.

A method for scaling small-scale wind tunnel studies to full scale was presented by Irwin et al. (2012) based on a 1:10 wind tunnel study and field observations of a full-scale roof paver failure. They observed in their wind tunnel experiments that pressure equalization is less effective in areas with high values of the surface pressure gradient. This is most likely because the vertical pressure gradient at the gaps that drives the pressure equalization flow is different from the pressure difference across other parts of the paver. Irwin et al. (2012) used a net uplift coefficient of $C_z = C_P/3$, where $C_P$ is the local pressure coefficient and the one-third scaling accounts for the pressure equalization. However, they noted that $C_P$ is sensitive to the Reynolds number and used the Reynolds number correction proposed by Hoxey et al. (2007) to scale the wind tunnel study results up to full scale. Their resulting full-scale prediction failure wind speed corresponded closely to the measured wind speed from a storm that produced roof paver failure. However, it should again be noted that the field data were based on post-storm observations rather than synchronous measurements of wind speed and failure.

## 3.4.2 Modular Green Roofs

Green roofs are becoming increasingly popular as roof ballast for their many environmental benefits. These include reduction in stormwater runoff (VanWoert et al. 2005), improved roof insulation (Castleton et al. 2010), reduced urban heat island effects (Susca et al. 2011), and improved aesthetics. This has led to green roof installations getting credit in various green development rating systems such as the U.S. Green Building Council's Leadership in Energy and Environmental Design (LEED) green building program (www.usgbc.org). Many green roof installations are modular, consisting of individual planter boxes delivered with soil and vegetation and ready to be laid out on the building's roof. There are, therefore, two sources of debris: (1) the plant and soil material and (2) the entire module. There is very little experimental data on the conditions for motion initiation for either of these two forms of failure.

A preliminary study was conducted at Southern Illinois University Edwardsville (Retzlaff et al. 2009) as part of a broader study of green roof behavior. A single green roof module was placed in a relatively small, high-speed aerodynamic wind tunnel and observed under a range of wind speeds. The authors also examined various treatments for inhibiting soil loss due to high wind shear. The study did not attempt to model the atmospheric boundary layer or the influence of the building on the wind field that would be experienced by an installed green roof module. As such, the results are largely of qualitative value. For example, it was shown that more fully vegetated modules increased the wind speed required to scour the growth media. Furthermore, various liquid binding agents were tested and some were shown to prevent scour even for un-vegetated modules.

A study by Wanielista et al. (2011) placed a full-scale un-vegetated green roof module on a 10 ft cube and exposed it to a Wall of Wind. The wind was at a 45° angle to the cube face and was initially set at 95 km/h (59 mi/h) for 30 s. When the wind speed was increased to 125 km/h (78 mi/h) the upwind corner of the module lifted up and testing was stopped. A parallel study of two full-scale installations was also conducted in which the local wind conditions and green roof performance were monitored for 9 months. No design wind events were recorded during the study.

A more extensive study of the impact of wind on green roofs was conducted by Prevatt et al. (2012). A series of tests were run using a Wall of Wind capable of generating speeds of up to 193 km/h (120 mi/h). A range of green roof modules with a variety of plants were placed on top of an 8-ft-high building and exposed to the wind. The authors present large amounts of visual data on the resulting plant damage as well as failure mechanisms for the modules as a whole. Again, a detailed parameterization of the conditions under which soil, plant, or module failure occurs was not presented. The report does, however, include a detailed literature review on the impact of wind on plants, soils, loose aggregate, and pavers.

The results of these and other tests were used to develop a method for calculating module motion initiation (Vo et al. 2013). Three motion initiation mechanisms were considered: sliding, uplift, and overturning. Sliding occurs when the pressure force acting on the vertical edges of an isolated module (or collection of modules) overcomes the friction force induced by the module weight. Uplift occurs when the pressure difference between the top and bottom of the module overcomes the module weight. In this case, no pressure equalization was accounted for in the calculations since the individual modules were often taped together for greater stability, thus preventing air flow between the modules (Cao et al. 2013). Overturning occurs when a net moment induced by the pressure distribution over the module overcomes the moment due to the module weight. In all cases, the modules are assumed to be rigid, although flexing is clearly visible in some of the photographs taken during the experiments of Prevatt et al. (2012).

Vo et al. (2013) present a worksheet for calculating the motion initiation criteria for a green roof system using ASCE 7-10 components and cladding loads assuming a 3 ft parapet around the building perimeter. No detailed parametric study results were presented to corroborate their model. Furthermore, soil scour is not accounted for in the worksheet although scour would act to reduce the module weight and lower the pressure coefficient required to initiate motion. Another attempt at quantification of failure criteria is presented by Ennis (2010). The model is based on ANSI/SPRI RP-14 (Single Ply Roofing Industry 2010). The model assumes that the soil in modules behaves as loose-laid aggregate and the modules behave like roof pavers. However, no test results were presented in support of these assumptions.

The only detailed parametric study focused on quantifying the conditions for motion initiation of green roof modules is presented by Cao et al. (2013). The authors ran a series of small-scale wind tunnel studies to find the lateral and uplift force coefficients as well as overturning moments for model green roof modules assuming no pressure equalization. The same three failure mechanisms identified by Vo et al. (2013) (sliding, uplift, and overturning) were considered. As with earlier studies, the modules are assumed to be held in place by their weight and behave as rigid bodies. The building parapet was found to be almost always beneficial, consistent with the results presented in Section 3.3.2. Increasing the area of the module, either through using larger modules or connecting sets of modules, reduced the net uplift. Surrounding modules with others reduced the lateral force coefficient. Their results suggest that ASCE 7-10 underestimates the uplift forces near the corners and edges of the roof and that these areas should not be used for green roof installation.

In all the studies conducted to date, no systematic parametric study of the conditions for motion initiation for the growth media, plants, modules, or collection of modules, has been conducted at full scale. Furthermore, most severe storms will be accompanied by rainfall, but the impact of soil moisture, and resulting increase in weight, has yet to be examined.

### 3.4.3 Roof Gravel

Roof gravel blow-off from built-up roofs is quite different from sand and aggregate motion initiation in open areas (as discussed in Section 3.2.2). On built-up roofs, the flow separation over the building and the resulting vortex structures create a highly heterogeneous unsteady flow field that strongly influences the aerodynamic forces acting on the aggregate particles. Therefore, the Shields diagram is inappropriate.

Extensive wind tunnel experiments were conducted over a range of scales by Kind (1974a, b, 1977) and Kind and Wardlaw (1977). These tests recreated the scour patterns observed in the field, including the V-shaped scour pattern caused by the delta wing vortex that forms from cornering flows. The results were used to develop a design guide for sizing parapets to protect roofs from gravel blow-off (Kind and Wardlaw 1976). This work forms the basis for current design standards, although some modifications have been proposed (Crandell and Fischer 2010).

Further wind tunnel studies at smaller scale were conducted by Kind (1986), in which it was shown that the small-scale tests did not match the earlier larger-scale experiments (Kind 1974a, b, 1977; Kind and Wardlaw 1977). This was attributed to particle Reynolds number effects as identified by Iversen and White (1982) and Shields (1936a, b, c). Kind (1986) proposed a Reynolds number correction method, based on the Shields diagram, to account for this effect.

However, Karimpour and Kaye (2012) showed that the correction, when applied to the Kind (1986) data, did not result in the small-scale data matching the larger-scale data. It is, therefore, possible that the original study from the 1970s may also have Reynolds number effects and thus have some limitations when applied at full scale.

This Reynolds number effect was also seen in a series of two-dimensional blow-off experiments done by Karimpour and Kaye (2012) using sand and millet seed as the model aggregate. The millet seed, although less dense than the sand, followed the same critical velocity behavior as the sand. For a wide rectangular building with the aggregate level with the top of the parapet (zero effective parapet height), Karimpour and Kaye (2012) found that the critical condition for aggregate blow-off was given by

$$K_c = 8.1 d_*^{-0.44} \tag{3.20}$$

where $K_c$ is the critical Tachikawa number based on the average velocity at the roof height. Values of $d_*$ up to 75 were tested but no Reynolds number–independent region was identified in the experimental results. Experiments with finite parapet heights showed similar qualitative behavior to the ballast paver tests of Bienkiewicz and Meroney (1988). For small parapet heights, when the critical Tachikawa number decreased the roof gravel became more susceptible to blow-off. For larger parapet heights the parapet increased the wind speed ($K$ value) required for motion initiation. This small parapet degradation in performance was not observed in Kind's wind tunnel studies.

### 3.4.4 Roof Sheathing

Roof sheathing panels can become airborne if the uplift force exceeds the restraining force of the fasteners used to affix them to the roof frame. The flight path of sheathing is very sensitive to the initial release conditions and the structure of the wake downstream from the building (Visscher and Kopp 2007). However, the conditions for lift-off are much more complex than a simple one-dimensional uplift/restraint balance. For larger sheathing panels there can be significant spatial variation in uplift that varies in time due to spatially small gusts. This results in time-varying loads in connections of individual sheathing to the roof frame (Prevatt et al. 2008) that can produce low-cycle fatigue or cycle-dependent effects in the connections (Guha and Kopp 2014). Furthermore, failure does not occur simultaneously across the panel so there can be significant off-loading that current static uplift tests do not address.

The need for dynamic testing of roofing components was illustrated by Baskaran et al. (2012). They showed that air intrusion between layers can reduce the net uplift on a given component, thus increasing the wind speed required for failure. It is, therefore, important to undertake dynamic testing with realistic wind pressure fluctuations and correctly installed full roof systems in order to accurately replicate the air flow between layers, and the resulting net uplift. A further discussion of risk-based modeling of roof sheathing liftoff is presented in Chapter 5.

### 3.4.5 Shingles

In general, roof shingles differ from roof sheathing in that they are smaller and are not secured around their entire perimeter. Instead, they are sealed along their top and again near their base but can be loose along their bottom (although there are a large variety of products with different geometry, materials, and fixing schedules). As with other roof coverings, motion is initiated when the uplift force overcomes the attachment force. However, because of the loose flap at the base of the shingle, the mechanism of uplift is different from that for perimeter-sealed roof coverings.

The major uplift forces on a shingle are due to stagnation pressure on the underside of the flap below the seal and the uplift due to low pressure over the top of the shingle (Peterka et al. 1997) (see Fig. 3.3). This stagnation pressure on the underside bends the flap upward into the wind field. This results in flow separation over the top of the flap, increasing the suction pressure on the upper side of the flap. It also increases the stagnation pressure on the underside (see Fig. 7 in Peterka et al. 1997) and the streamwise projected area of the shingle (see Fig. 3.4). As such, wind uplift of shingles occurs predominantly on the windward side of buildings with relatively steep roofs (Marshall et al. 2010).

A model for shingle uplift, presented by Peterka et al. (1997), is based on measured pressure coefficients on each side of the shingle and the gust wind speed over the surface of the roof. Peterka et al. (1997) measured pressure coefficients of $\Delta C_p = -0.4$ for the seal strip to the front edge of the shingle, $\Delta C_p = -0.1$ from the top of the cutout to the seal strip, and $\Delta C_p = -0.4$ from the top of the cutout to the front edge for unsealed shingles. Peterka et al. (1997) also made measurements of the wind speed over the surface of the roof and found that the worst case was a surface wind speed 1.6 times the approach wind speed at the eave height. Therefore, the uplift pressure is given by

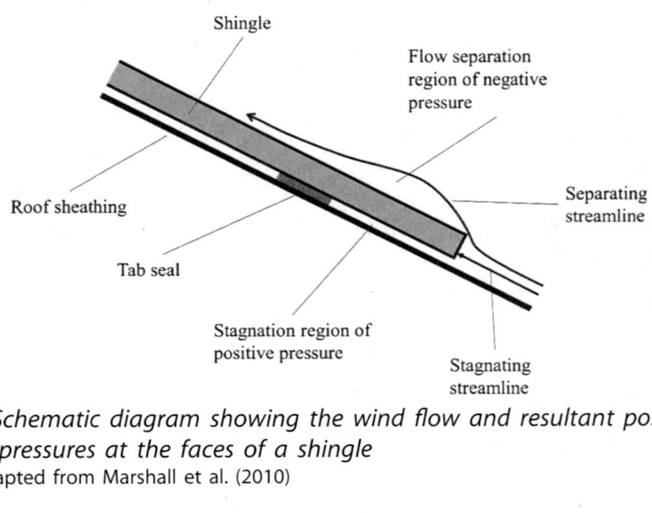


*Fig. 3.3. Schematic diagram showing the wind flow and resultant positive and negative pressures at the faces of a shingle*

Source: Adapted from Marshall et al. (2010)

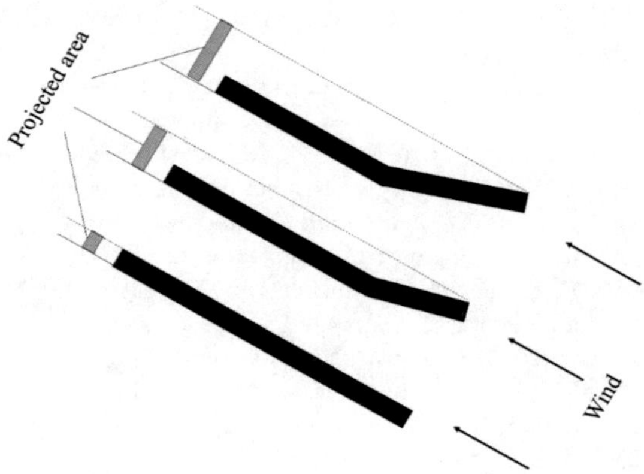


*Fig. 3.4. Schematic diagram showing the increase in the streamwise projected area (gray line) as the shingle is uplifted in the wind*
Source: Adapted from Marshall et al. (2010)

$$P = \frac{1}{2}\rho U_{\text{eave}}^2 2.5DC_p \tag{3.21}$$

The value of 2.5 ($= 1.6^2$) for the dynamic pressure correction due to wind speed up over the roof surface was confirmed by the measurements of Dixon et al. (2012).

## 3.5 CONCLUSIONS

There is extensive literature on debris motion initiation for a broad range of debris types. The literature contains a lot of empirical measurements of motion initiation conditions supported by theoretical models. However, caution should be exercised since there is still considerable variability in many aspects of debris motion initiation. The two main areas of variability are in the fixing forces and the wind field. For example, there is a broad range of roof shingles on the market and variability in how well they are installed, and therefore in their ability to resist wind uplift forces. There is also significant variability in the wind flow over the roof of a building, which translates into uplift variability on the shingles.

### *References*

Aly, A. M., Bitsuamlak, G. T., and Chowdhury, A. G. (2012). "Full-scale aerodynamic testing of a loose concrete roof paver system." *Eng. Struct.*, **44**, 260–270.

ASCE. (2010). "Minimum design loads for buildings and other structures." *ASCE/SEI 7-10*, Reston, VA.

Bagnold, R. A. (1937). "The transport of sand by wind." *Geograph. J.*, **89**(5), 409–438.

Bagnold, R. A. (1941). *The physics of blown sand and desert dunes*, Methuen, London.

Baker, C. J., and Bell, H. J. (1992). "The aerodynamics of urban trees." *J. Wind Eng. Ind. Aerodyn.*, **44**(1–3), 2655–2666.

Banks, D. (2000). "The suction induced by conical vortices on low-rise buildings with flat roofs." Ph.D. thesis, Colorado State Univ., Fort Collins, CO.

Banks, D., Ghua, T. K., and Fewless, Y. J. (2015). "A hybrid method of generating realistic full-scale time series of wind loads from large-scale wind tunnel studies: Application to solar arrays." *Proc., 14th Int. Conf. on Wind Engineering*, Porto Alegre, Brazil, International Association for Wind Engineering, Kanagawa, Japan.

Baskaran, A., Molleti, S., Ko, S., and Shoemaker, L. (2012). "Wind uplift performance of composite metal roof assemblies." *J. Archit. Eng.*, **18**(1), 2–15.

Beck, V., and Dotsek, N. (2010) "Reconstruction of near-surface tornado wind fields from forest damage." *J. Appl. Meteorol. Climatol.*, **49**(7), 1517–1537.

Bienkiewicz, B., and Meroney, R. (1988). "Wind effects on roof ballast pavers." *J. Struct. Eng.*, **114**(6), 1250–1267.

Bienkiewicz, B., and Sun, Y. (1992). "Wind-tunnel study of wind loading on loose-laid roofing systems." *J. Wind Eng. Ind. Aerodyn.*, **43**(1–3), 1817–1828.

Bienkiewicz, B., and Sun, Y. (1997). "Wind loading and resistance of loose-laid roof paver systems." *J. Wind Eng. Ind. Aerodyn.*, **72**, 401–410.

Cao, J., Tamura, Y., and Yoshida, A. (2013). "Wind tunnel investigation of wind loads on rooftop model modules for green roofing systems." *J. Wind Eng. Ind. Aerodyn.*, **118**, 20–34.

Castleton, H. F., Stovin, V., Beck, S. B. M., and Davison, J. B. (2010). "Green roofs: Building energy savings and the potential for retrofit." *Energy Build.*, **42**(10), 1582–1591.

Cochran, L. S., and English, E. C. (1997). "Reduction of roof wind loads by architectural features." *Archit. Sci. Rev.*, **40**(3), 79–87.

Crandell, J., and Fischer, M. (2010). "Winds of change: Resolving roof aggregate blow-off." *Proc., RCI 25th Int. Convention and Trade Show*, RCI, Raleigh, NC.

Cullen, S. (2002). "Trees and wind: Wind scales and speeds." *J. Arboriculture*, **28**(5), 237–242.

Dixon, C. R., Masters, F. J., Prevatt, D. O., and Gurley, K. R. (2012). "Investigation of the wind resistance of asphalt shingles." *Advances in hurricane engineering: Learning from our past*, ASCE, Reston, VA, 517–527.

Ennis, M. (2010). "Fire and wind resistance standards for vegetative roofs." *Proc., RCI 25th Int. Convention and Trade Show*, RCI, Raleigh, NC.

Guha, T. K., and Kopp, G. A. (2014). "Storm duration effects on roof-to-wall-connection failures of a residential, wood-frame, gable roof." *J. Wind Eng. Ind. Aerodyn.*, **133**, 101–109.

Guo, J. (2002). "Hunter Rouse and Shields diagram." *Proc., 13th IAHR-APD Congress on Advances in Hydraulics and Water Engineering*, Vol. **2**, World Scientific, Singapore, 1096–1098.

Holmes, J. D., Baker, C. J., and Tamura, Y. (2006). "Tachikawa number: A proposal." *J. Wind Eng. Ind. Aerodyn.*, **94**(1), 41–47.

Hölscher, N., and Niemann, H. (1998) "Towards quality assurance for wind tunnel tests: A comparative testing program of the Windtechnologische Gesellschaft." *J. Wind Eng. Ind. Aerodyn.*, **74–76**, 599–608.

Hoxey, R., Richards, P., Castro, I., and Lim, H. C. (2007). "Reynolds number effects on bluff bodies in the atmospheric boundary layer: A review of the full-scale measurements made at Silsoe." *Proc., 12th Int. Conf. of Wind Engineering*, Cairns, Australia, International Association for Wind Engineering, Kanagawa, Japan.

Irwin, P., Dragoiescu, C., Cicci, M., and Thompson, G. (2012). "Wind tunnel model studies of aerodynamic lifting of roof pavers." *Advances in hurricane engineering: Learning from our past*, ASCE, Reston, VA, 496–505.

Iversen, J. D., and White, B. R. (1982). "Saltation threshold on Earth, Mars and Venus." *Sedimentology*, **29**(1), 111–119.

James, K. R., Haritos, N., and Ades, P. K. (2006). "Mechanical stability of trees under dynamic loads." *Am. J. Botany*, **93**(10), 1522–1530.

Kane, B., and Smiley, E. T. (2006). "Drag coefficient and crown area estimation for red maple." *Can. J. For. Res.*, **36**(8), 1951–1958.

Karimpour, A., and Kaye, N. B. (2012). "The critical velocity for aggregate blow-off from a built-up roof." *J. Wind Eng. Ind. Aerodyn.*, **107–108**, 83–93.

Kind, R. J. (1974a). "Estimation of critical wind speeds for scouring of gravel or crushed stone on rooftops." *LTR-LA-142*, National Research Council of Canada, Ottawa, Canada.

Kind, R. J. (1974b). "Wind tunnel tests on some building models to measure wind speeds at which is gravel is blown off rooftops." *LTR- LA-146*, National Research Council of Canada, Ottawa, Canada.

Kind, R. J. (1977). "Further wind-tunnel tests on building models to measure wind speeds at which gravel is blown off rooftops." *LTR-LA-189*, National Research Council of Canada, Ottawa, Canada.

Kind, R. J. (1986). "Measurement in small wind tunnels of wind speeds for gravel scour and blow off from rooftops." *J. Wind Eng. Ind. Aerodyn.*, **23**, 223–235.

Kind, R. J., and Wardlaw, R. L. (1976). *Design of rooftops against gravel blow-off*, National Research Council of Canada, Ottawa, Canada.

Kind, R. J., and Wardlaw, R. L. (1977). "Development of a procedure for the design of rooftops against gravel blow-off and scour in high winds." *Proc., Symp. on Roofing Technology*, National Roofing Contractors Association, Chicago, 112–123.

Kopp, G. A., Surry, D., and Mans, C. (2005). "Wind effects of parapets on low buildings. 1: Basic aerodynamics and local loads." *J. Wind Eng. Ind. Aerodyn.*, **93**(11), 817–841.

Lim, H. C., Catro, I. P., and Hoxey, R. P. (2007). "Bluff bodies in deep turbulent boundary layers: Reynolds-number issues." *J. Fluid Mech.*, **571**, 97–118.

Marshall, T. P., Morrison, S. J., Herzog, R. F., and Green, J. R. (2010). "Wind effects on asphalt shingles." *Proc., 29th Conf. on Hurricanes and Tropical Meteorology*, Tucson, AZ, American Meteorological Society, Boston, MA.

Mooneghi, M. A., Irwin, P., and Chowdhury, A. G. (2015). "Partial turbulence simulation method for small structures." *Proc., 14th Int. Conf. on Wind Engineering*, Porto Alegre, Brazil, International Association for Wind Engineering, Kanagawa, Japan.

Peltola, H., and Kellomaki, S. (1993). "A mechanistic model for calculating windthrow and stem breakage of Scots pines at stand edge." *Silva Fenn.*, **27**(2), 99–111.

Peterka, J. A., et al. (1997). "Wind uplift model for asphalt shingles." *J. Archit. Eng.*, **3**(4), 147–155.

Prevatt, D. O., Acomb, G. A., Masters, F. J., Vo, T. D., and Schild, N. K. (2012). "Comprehensive wind uplift study of modular and built-in-place green roof systems." *Rep. No.UF01-12*, Univ. of Florida, Gainesville, FL.

Prevatt, D. O., Schiff, S. D., Stamm, J. S., and Kulkarni, A. S. (2008). "Wind uplift behavior of mechanically attached single-ply roofing systems: The need for correction factors in standardized tests." *J. Struct. Eng.*, **134**(3), 489–498.

Putz, F. E., Coley, P. D., Lu, K., Montalvo, A., and Aiello, A. (1983). "Uprooting and snapping of trees: Structural determinants and ecological consequences." *Can. J. For. Res.*, **13**(5), 1011–1020.

Retzlaff, B., Celik, S., Morgan, S., Luckett, K., and Graham, M. (2009). "Wind uplift of green roof systems." *Final Rep.*, Southern Illinois Univ. Edwardsville, Edwardsville, IL.

Richards, P. J., Hoxey, R. P., and Short, L. J. (2001). "Wind pressures on a 6 m cube." *J. Wind Eng. Ind. Aerodyn.*, **89**(14–15), 1553–1564.

Shields, A. (1936a). "Anwendung der aehnlichkeitsmechanik und der turbulenzforschung auf die geschiebebewegung." Ph.D. dissertation, Technischen Hochschule, Berlin (in German).

Shields, A. (1936b). "Anwendung der aehnlichkeitsmechanik und der turbulenzforschung auf die geschiebebewegung." Mitteilungen der Preussischen Versuchsanstalt fur Wasserbau und Schiffbau, Berlin (in German).

Shields, A. (1936c). *Application of similarity principles and turbulence research to bed-load movement*, Trans. W. P. Ott and J. C. van Uchelen, California Institute of Technology, Pasadena, CA.

Shigo, A. L. (1985). "How tree branches are attached to trunks." *Can. J. Botany*, **63**(8), 1391–1401.

Single Ply Roofing Industry. (2010). "Wind design standard for ballasted single-ply roofing systems." *ANSI/SPRI RP-14*, Waltham, MA.

Slater, D., and Harbinson, C. (2010). "Towards a new model of branch attachment." *Arboriculture J.*, **33**(2), 95–105.

Stathopoulos, T., and Baskaran, A. (1987). "Wind pressures on flat roofs with parapets." *J. Struct. Eng.*, **113**(11), 2166–2180.

Stathopoulos, T., Marathe, R., and Wu, H. (1999). "Mean wind pressures on flat roof corners affected by parapets: Field and wind tunnel studies." *Eng. Struct.*, **21**(7), 629–638.

Suaris, W., and Irwin, P. (2010). "Effect of roof-edge parapets on mitigating extreme roof suctions." *J. Wind Eng. Ind. Aerodyn.*, **98**(10–11), 483–491.

Sun, Y. (1993). "Wind loading on loose-laid roofing paver systems." Ph.D. dissertation, Colorado State Univ., Fort Collins, CO.

Susca, T., Gaffin, S. R., and Dell'Osso, G. R. (2011). "Positive effects of vegetation: Urban heat island and green roofs." *Environ. Pollut.*, **159**(8–9), 2119–2126.

Tieleman, H. W., Ge, Z., Hajj, M. R., and Reinhold, T. A. (2003) "Pressures on a surface-mounted rectangular prism under varying incident turbulence" *J. Wind Eng. Ind. Aerodyn.*, **91**(9), 1095–1115.

VanWoert, N. D., Rowe, D. B., Andresen, J. A., Rugh, C. L., Fernandez, R. T., and Xiao, L. (2005). "Green roof stormwater retention." *J. Environ. Q.*, **34**(3), 1036–1044.

Visscher, B. T., and Kopp, G. A. (2007). "Trajectories of roof sheathing panels under high winds." *J. Wind Eng. Ind. Aerodyn.*, **95**(8), 697–713.

Vo, T. D., Prevatt, D. O., Acomb, G. A., and Masters, F. J. (2013). "A preliminary approach to determining component and cladding wind loads on modular green roof systems." *Proc., 12th Americas Conference on Wind Engineering (12ACWE)*, Seattle, American Association for Wind Engineering, Fort Collins, CO.

Wanielista, M., Minareci, M., Catbas, N., and Hardin, M. (2011). "Green roofs and wind loading." *FDEP Project No. WM 934*, Univ. of Central Florida, Orlando, FL.

Wills, J. A. B., Lee, B. E., and Wyatt, T. A. (2002). "A model of wind-borne debris damage." *J. Wind Eng. Ind. Aerodyn.*, **90**(4–5), 555–565.

Wilson, C. A. M. E., et al. (2010). "The hydrodynamic drag of full scale trees." *Proc., Int. Conf. on River Flow*, A. Dittrich, K. Koll, J. Aberle, and P. Geisenhainer, eds., Vol. **1**, Bundesanstalt für Wasserbau, Karlsruhe, Germany, 453–459.

# CHAPTER 4

# Flight Characteristics of Wind-Borne Debris

Gregory A. Kopp*

In this chapter the equations of motion for the flight of wind-borne debris are discussed, with a focus on flight trajectories and debris speeds for the three classes of debris. The effects of turbulence and the local flow field around building surfaces are emphasized.

## 4.1 EQUATIONS OF MOTION FOR WIND-BORNE DEBRIS

### 4.1.1 Definitions and Equations for Two-Dimensional Motion

The flight of wind-borne debris is governed by Newton's second law of motion

$$\vec{F} = m\vec{a} = m\frac{d^2\vec{x}}{dt^2} \tag{4.1}$$

where $\vec{F}$ is the force applied to the object of mass, $m$; $\vec{x}$ is the location of the object in space at time $t$; and $\vec{a}$ is the acceleration of the object.

For two-dimensional motion, the forces acting on the flying object are the drag, $F_D$, lift, $F_L$, and torque, $M$, in a two-dimensional coordinate system $(x, z)$, as depicted in Fig. 4.1 with the relative wind speed, $V_{\text{rel}}$, acting on the object. The angle to the horizontal, $\beta$, and the angle of attack, $\xi$, are also defined in Fig. 4.1. Applying these to Eq. (4.1), one obtains

$$\begin{aligned}
m\frac{d^2x}{dt^2} &= m\frac{du}{dt} = 0.5\rho A(u_{\text{rel}}^2 + v_{\text{rel}}^2)\{C_D\cos(\beta) - C_L\ \sin(\beta)\} \\
m\frac{d^2z}{dt^2} &= m\frac{dv}{dt} = 0.5\rho A(u_{\text{rel}}^2 + v_{\text{rel}}^2)\{C_D\sin(\beta) + C_L\ \cos(\beta)\} - mg \\
I\frac{d^2\theta}{dt^2} &= 0.5\rho AL(u_{\text{rel}}^2 + v_{\text{rel}}^2)C_M
\end{aligned} \tag{4.2}$$

---

*Boundary Layer Wind Tunnel Laboratory, Faculty of Engineering University of Western Ontario, London, ON Canada.

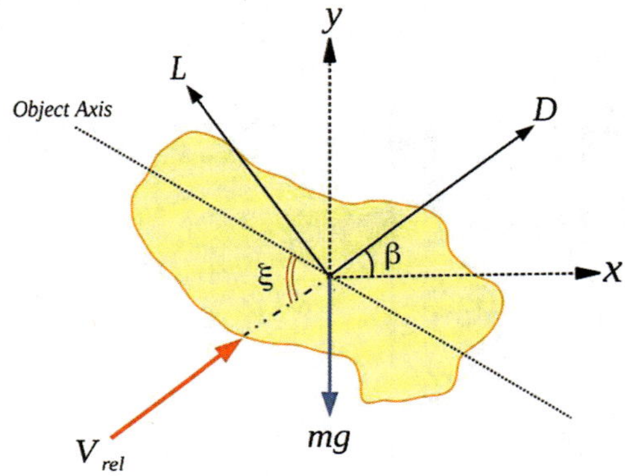


*Fig. 4.1. Definition sketch of forces acting on a wind-borne object*

where $A$ is the area of the object; $L$ is the length of the object; $I$ is the mass moment of inertia; $\theta$ is the rotational position; $C_D = F_D/0.5\rho AV_{\text{rel}}^2$ is the drag coefficient; $C_L = F_L/0.5\rho AV_{\text{rel}}^2$ is the lift coefficient; $C_M = M/0.5\rho AlV_{\text{rel}}^2$ is the moment coefficient; $\rho$ is the air density; and $g$ is the gravitational constant. The relative wind speed, $V_{\text{rel}}$, is defined as

$$V_{\text{rel}}^2 = (U - u)^2 + (V - v)^2 = u_{\text{rel}}^2 + v_{\text{rel}}^2 \tag{4.3}$$

where $U$ is the horizontal component of the instantaneous wind speed; $V$ is the vertical component of the instantaneous wind speed; $u$ is the horizontal component of the object's instantaneous speed; $v$ is the vertical component of the object's speed; and $u_{\text{rel}}$ and $v_{\text{rel}}$ are the horizontal and vertical components of the relative wind speed, respectively. Later, we will consider the wind flow around buildings, where we will use the mean wind speed at the mean roof height, $U_H$, and the peak, 3-s gust speed at the mean roof height, $\hat{U}_H$.

### 4.1.2 Normalized Equations of Motion

It is often convenient to use non-dimensional forms for the equations of motion given by Eq. (4.2). Here, we normalize the spatial and temporal coordinates as

$$\bar{x} = \frac{xg}{\hat{U}_H^2},\ \bar{z} = \frac{zg}{\hat{U}_H^2},\ \bar{t} = \frac{tg}{\hat{U}_H} \tag{4.4}$$

where the overbar indicates a non-dimensionalized variable. This approach is similar to that used by Tachikawa (1983), who studied the flight of debris-like objects in uniform, smooth flow. Because real debris flies in the highly turbulent flow around buildings in the atmospheric surface layer, we normalize the spatial coordinates using a wind speed of relevance to design. In the American wind loading standard, ASCE 7-10 (ASCE 2010), the peak, 3-s gust speed at the mean roof height, $\hat{U}_H$, is used for design, so we use this in the normalization here. Using these non-dimensional variables, Eq. (4.2) can be written as

$$\frac{d^2\overline{x}}{d\overline{t}^2}=\frac{d\overline{u}}{d\overline{t}}=K(\overline{u}_{\text{rel}}^2+\overline{v}_{\text{rel}}^2)\{C_D\cos(\beta)-C_L\ \sin(\beta)\}$$
$$\frac{d^2\overline{z}}{d\overline{t}^2}=\frac{d\overline{v}}{d\overline{t}}=K(\overline{u}_{\text{rel}}^2+\overline{v}_{\text{rel}}^2)\{C_D\sin(\beta)+C_L\ \cos(\beta)\}-1$$
$$\frac{d^2\overline{\theta}}{d\overline{t}^2}=K\ \Delta\ \text{Fr}^2\ (\overline{u}_{\text{rel}}^2+\overline{v}_{\text{rel}}^2)\ C_M \tag{4.5}$$

where

$$\overline{u}_{\text{rel}}=\frac{u_{\text{rel}}}{\hat{U}_H},\ \overline{v}_{\text{rel}}=\frac{v_{\text{rel}}}{\hat{U}_H},\ \overline{u}=\frac{u}{\hat{U}_H},\ \overline{v}=\frac{v}{\hat{U}_H} \tag{4.6}$$

Eq. (4.7) is the Tachikawa number (Holmes et al. 2006)

$$K=\frac{0.5\rho\hat{U}_H^2A}{mg}, \tag{4.7}$$

$$\Delta=\frac{ml^2}{I}, \tag{4.8}$$

and Eq. (4.9) is the Froude number

$$\text{Fr}=\frac{\hat{U}_H}{\sqrt{gl}} \tag{4.9}$$

Eq. (4.5) can be considered as exact for two-dimensional motion presuming the aerodynamic force and moment coefficients are known instantaneously. In most practical applications, models for these aerodynamic coefficients are employed that assume quasi-steady theory. Although such models are known to have some limitations (e.g., Kordi and Kopp 2009), these are expected to be sufficiently accurate for wind-borne debris applications where other parameters have greater uncertainty, as discussed below.

As discussed in Chapter 2, debris is typically categorized as compact, rod-like, or plate-like (Wills et al. 2002) based on the aerodynamic characteristics. Compact debris is assumed to be one-dimensional, such as gravel, which can be approximated by spheres. Rod-like debris, like 2 × 4 lumber, is approximately two-dimensional with one long dimension. Plate-like debris is like two-dimensional sheets (with a length and a width) with one thin dimension (thickness), such as asphalt shingles or plywood sheathing. Each of these has distinct aerodynamic characteristics.

### 4.1.3 Compact Debris

For compact debris it is usual to assume that the lift and moment coefficients are negligible so that the only aerodynamic force coefficient is the drag. This simplifies the equations to

$$\frac{d^2\overline{x}}{d\overline{t}^2} = \frac{d\overline{u}}{d\overline{t}} = K(\overline{u}_{\text{rel}}^2 + \overline{v}_{\text{rel}}^2)\{C_D \cos(\beta)\}$$
$$\frac{d^2\overline{z}}{d\overline{t}^2} = \frac{d\overline{v}}{d\overline{t}} = K(\overline{u}_{\text{rel}}^2 + \overline{v}_{\text{rel}}^2)\{C_D \sin(\beta)\} - 1 \tag{4.10}$$

Even though the equations in Eq. (4.10) are relatively simple, there is no analytical solution and the equations must be solved numerically. However, if the vertical air resistance is neglected, there is an analytical solution, as shown by Holmes (2004), but which underestimates the horizontal speeds and flights distances (Letchford et al. 2013).

For spheres in a horizontal, uniform, smooth flow, the non-dimensional asymptotic horizontal speed, obtained from Eq. (4.10) assuming negligible lift and moment coefficients, is $\overline{u} = 1$, while the vertical speed is $\overline{v} = -(1/\sqrt{KC_D})$ (Baker 2007). When the Tachikawa number is relatively large (i.e., for high wind speeds or a small object weight), $\overline{v}$ is small and the debris stays airborne for relatively longer distances.

### 4.1.4 Plate-Like Debris

The aerodynamic coefficients for plates usually use quasi-steady models that include static force coefficients, which depend on the angle of attack, $\zeta$, and autorotational coefficients, which depend on the rotational speed, $d\theta/dt$, of the object. Such models can be found in Tachikawa (1983), Baker (2007), Lin et al. (2007), and Kordi and Kopp (2009), among others. Using these models, the trajectory can be solved if the wind field is known or assumed.

In steady flow, debris speeds can reach an asymptotic (steady) limit. By averaging over one cycle of rotation and setting $du/dt = 0$, the non-dimensional asymptotic horizontal speed for thin plates in uniform, smooth flow is of the following form (Baker 2007):

$$\overline{u} = 1 \pm C_a\sqrt{1/K} \tag{4.11}$$

For thin square plates, $C_a = 0.26$, while for thin plates with a 2:1 aspect ratio (such as plywood sheathing), $C_a = 0.42$ (Kordi and Kopp 2009). Thus, it can be seen that debris could travel at speeds above that of the wind in theory. In practice, because of the direction of plate rotation, it is the negative sign that is typically selected in Eq. (4.11) such that the asymptotic limit is $\overline{u} \leq 1$, although overspeeding has been observed in experiments (Lin et al. 2006).

## 4.2 FLIGHT TRAJECTORIES

### 4.2.1 Initial Motion—Plate-Like Debris

Wills et al. (2002) introduced the concept of the "fixture strength integrity," or the resistance of the building component, to the analysis of debris motion, as discussed in Chapter 3. The fixture strength integrity, $\phi$, is simply the ratio of the force required to break the object loose to its mass. These authors pointed out that loose-laid objects tend to tumble along the ground or building surface. This is caused by the fact that as an object begins to accelerate, the magnitude of the relative velocity decreases, reducing the aerodynamic forces (which depend on $V_{rel}^2$) and allowing the object to fall back to the ground. This process can be observed with the scour of gravel on a roof surface, for example. In contrast, building components with large hold-down forces will fail at relatively high wind speeds, compared to loose-laid objects, and will tend to stay airborne for much longer distances.

The initial condition for flight is defined as

$$\frac{d^2\overline{x}}{d\overline{t}^2}=K\{C_D\cos(\alpha)-C_L\ \sin(\alpha)\}$$
$$\frac{d^2\overline{z}}{d\overline{t}^2}=K\{C_D\sin(\alpha)+C_L\ \cos(\alpha)\}-1$$
$$\frac{d^2\overline{\theta}}{d\overline{t}^2}=K\ \Delta\ \mathrm{Fr}^2\ C_M \tag{4.12}$$

because the relative wind speed is equal to the actual wind speed at failure, where α is the angle of the object with respect to the horizontal on the building surface. This can be simplified for an object fastened to a nearly flat roof, i.e.

$$\frac{d^2\overline{z}}{d\overline{t}^2}=KC_L-1=\phi-1 \tag{4.13}$$

By including the fixture strength integrity, one can see that $\phi$ defines the initial acceleration and controls the initial trajectory. Of course, the building aerodynamics control $C_L$, and therefore the wind speed at which failure occurs. As shown by Letchford et al. (2013), the solution to Eq. (4.13) at the onset of motion is

$$\overline{v}=(KC_L-1)\,\overline{t}=(\phi-1)\,\overline{t}\quad\text{and}$$
$$\overline{z}=0.5(KC_L-1)\,\overline{t}^2=0.5(\phi-1)\,\overline{t}^2 \tag{4.14}$$

For larger $\phi$, the initial motion would cause the component to go higher, potentially into higher speed flow when one considers the nature of flow above the roof. This could potentially lead to disproportionately longer trajectories when compared to components with small $\phi$, the latter of which may not leave the roof surface depending on the precise details of the flow field.

The most vulnerable locations for components and cladding on a building surface tend to be in regions with high suction. These are usually associated with building-generated vortices and flow separations. These flow patterns have a significant impact on the flight trajectories of debris. Fig. 4.2 shows the initial flight patterns for (model) shingles on a roof in experiments conducted by Kordi and Kopp (2011). Fig. 4.2 shows that there can be significant variability in the initial flow patterns because of the variations in the flow field above the roof surface at different locations on the roof and also variations in the flow patterns at the

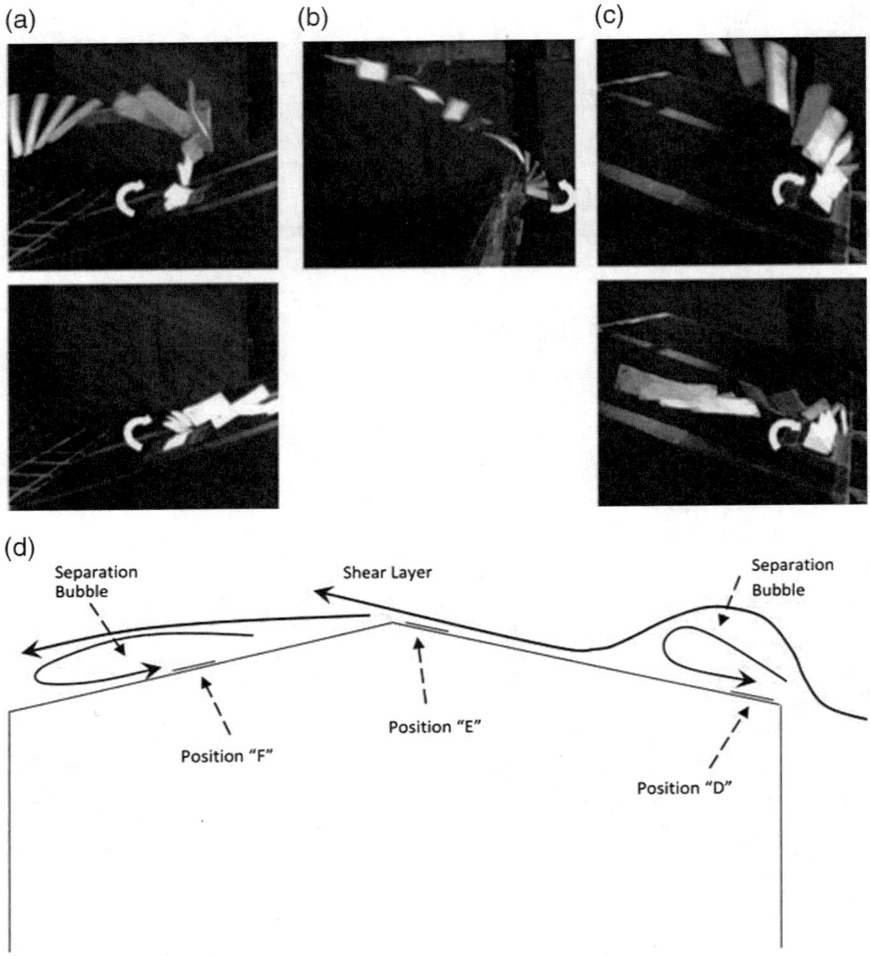


*Fig. 4.2. Flight patterns of shingles from a gable roof at three locations: (a) leeward side of the ridge (Position F); (b) windward side of the ridge (Position E); (c) windward eave (Position D); (d) schematic of the flow patterns over the roof for wind from right to left*

Source: From Kordi and Kopp (2011)

moment of failure. Of particular importance appear to be the patterns associated with separation "bubbles," where the local flow direction is opposite to the bulk wind direction close to the building surface. When the component can clear the separation bubble, it enters the accelerated flow above the separation bubble and can travel a substantial distance. Components that do not enter this high-speed flow tend to stay on the roof surface. Upstream terrain and surrounding buildings can substantially alter these flow patterns and the subsequent debris trajectories.

### 4.2.2 Initial Motion—Compact Debris

The initial motion of roof gravel is along the roof surface and is caused by the wind field near the roof surface, i.e., the flow patterns generated by the building. Initially, only a few rocks will be moving, but as the wind speed increases the gravel begins to scour, as shown in Fig. 4.3. This is consistent with the effects of the fixture strength integrity suggested by Wills et al. (2002), where loose-laid objects

(a)

(b)

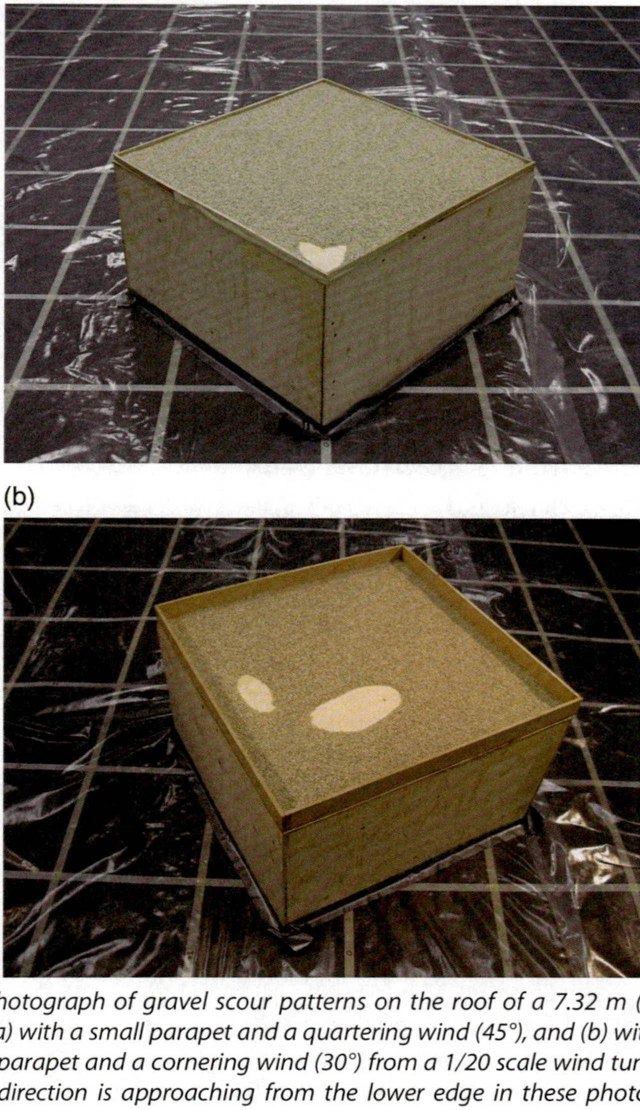

*Fig. 4.3. Photograph of gravel scour patterns on the roof of a 7.32 m (24 ft) high building: (a) with a small parapet and a quartering wind (45°), and (b) with a 0.91 m (3 ft) high parapet and a cornering wind (30°) from a 1/20 scale wind tunnel model. The wind direction is approaching from the lower edge in these photographs*

tend to move primarily along the surface. Kind (1974a, b) found that the onset speed for a particular building size and shape is proportional to $\sqrt{d}$, where $d$ is the diameter of the gravel. Scour (and flight) speeds also depend on the height of the building, parapet height, surrounding buildings, and wind direction, which alter the wind speeds and patterns near the roof surface. At wind speeds above the scour speed, gravel ballast will leave the roof, typically leaving the roof surface over the windward parapet first, and then, at higher speeds, above the leeward parapet.

Because spherical objects do not tend to generate lift, a vertical component of the wind is required to lift them off the roof surface. Such vertical motions are generated by the flow separations along the roof edges and corners. The so-called corner vortices, discussed in Chapter 3, which are generated by oblique wind directions, induce particularly strong upward motions. Unlike roof components, which are loaded by the pressure fields associated with these vortices, the loads on roof gravel depend on the wind speeds and directions generated by these vortices. One outcome of this is that the dependence on wind direction is particularly significant for gravel motion with the wind speeds required for onset of flight for wind directions normal to the wall being more than double those for quartering winds. In other words, directionality effects are larger for roof gravel than for typical roof components. Quartering winds tend to yield the lowest flight speeds because of the strong vertical flow components associated with cornering vortices. High parapets can substantially increase the onset of flight gust wind speeds, although for typical low-rise buildings, the onset of flight is at wind speeds that are usually below 161 km/h (100 mi/h) for the critical wind directions (Masters and Gurley 2008).

### 4.2.3 Flight Patterns—Plate-Like Debris

Many of the experiments and analyses on flight of wind-borne debris have focused on uniform, smooth flow with well-defined initial conditions such that the initial forces and moments at release are closely related to the static aerodynamic coefficients. Tachikawa (1983) was the first to do this, but much of the subsequent work has followed this same approach. Such experiments have provided validation for numerical models, debris flight speeds to set impact test criteria (e.g., Lin et al. 2007), and flight trajectories for risk models (e.g., Tachikawa 1988). For plate-like debris, these experiments have shown that the initial conditions have a significant effect on flight trajectories. Depending on the initial angle of attack, $\xi$, and the Tachikawa number, $K$, different modes of flight can arise, including cases with (1) significant rotation, (2) mostly translational motion with limited rotation, and (3) reversals of the direction of rotation mid-flight. Fig. 4.4 shows sample trajectories taken from the Tachikawa (1983) seminal work on the two-dimensional motion of plates. These patterns call to mind the patterns of falling paper, the only difference being that falling paper typically is not placed in cross-flow, which is the driving force for wind-borne debris. Visscher and Kopp (2007) have shown these same patterns can arise for debris originating from the roof of a building, with the pattern significantly affecting the total distance the debris flies. Tachikawa (1988) used these significant effects of initial conditions to estimate variability of trajectories in real storms, an approach that has been used many times subsequently, as discussed in

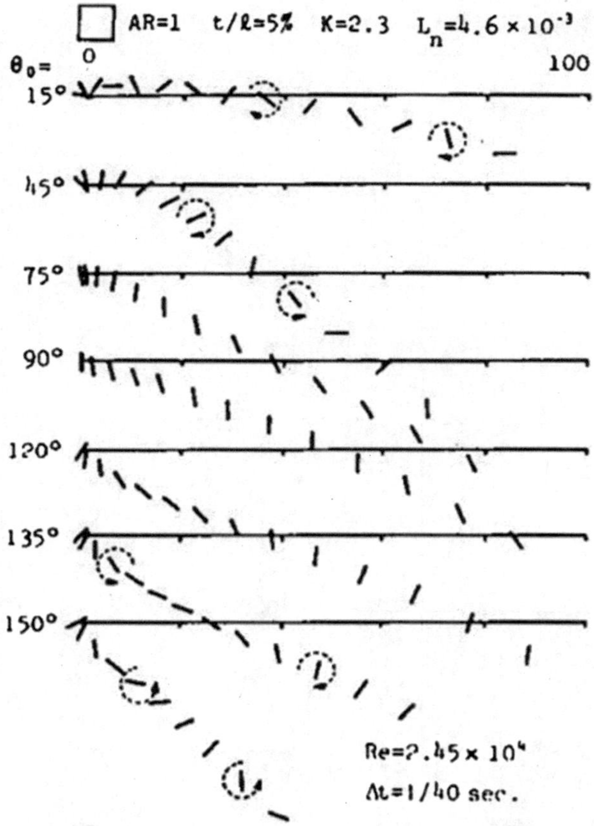


*Fig. 4.4. Example of typical trajectories of thin plates in uniform smooth flow for different initial angles of attack*
Source: From Tachikawa (1983)

Chapter 6. However, the flow fields near building surfaces and in building wakes, and the source location on the building surface, often play larger roles in determining the actual trajectories (Kordi et al. 2010; Kordi and Kopp 2011).

## 4.2.4 Overall Flight Distances and Trajectories—Plate-Like Debris

Overall flight distances and trajectories are important for establishing the risk of debris impact. Trajectories depend on the wind speed and direction; the mass and shape of the object; the original location of the component on the source building; the size, shape, and parapet heights of the source building, and the location and height of any adjacent buildings. As shown by Fig. 4.4, there is inherent variability in flight patterns in uniform smooth flow. The variation in the turbulent flow patterns around buildings also causes substantial variation in the subsequent debris trajectory, as indicated by Fig. 4.2. For example, Fig. 4.5 shows the landing locations of shingles originating from different locations on the roof of a house.

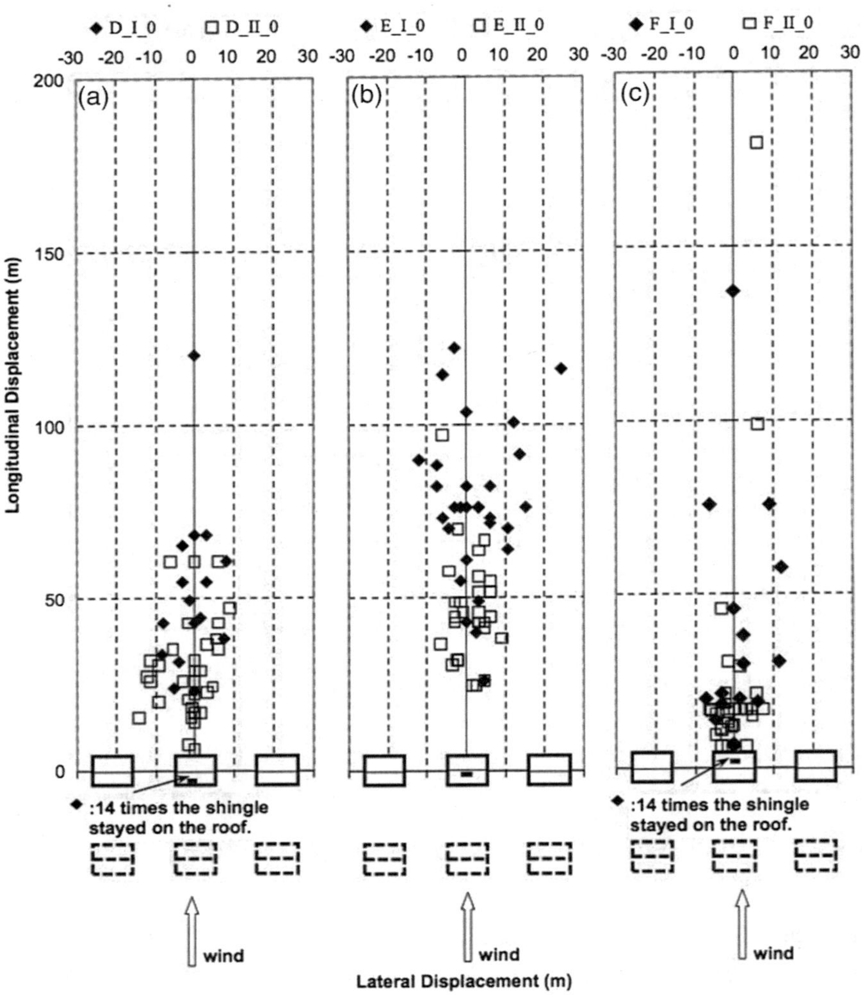


*Fig. 4.5. Landing locations for shingles originating from the (a) windward eave, (b) windward side of the ridge, and (c) leeward side of the ridge for a gable-roofed house with (open symbols) and without (closed symbols) surrounding houses*
Source: From Kordi and Kopp (2011)

There is clearly a wide variation of landing locations under nominally identical initial conditions, with one source roof position having a range from 0 to ~180 m (0 to ~590 ft) in this particular case. Kordi and Kopp (2011) attributed this primarily to variations in the turbulent velocity field at the location of the shingle at the moment of failure, as well as to variations during flight.

Uniform flow results suggest that the higher the wind speed, the further the flight. However, for debris originating on real buildings, the local flow patterns around the position of the component at the instant of failure play a larger role, altering the flight distances. As a result, numerical solutions to Eq. (4.5) that utilize uniform flow will not always be accurate for modeling flight trajectories because the local variations in wind speed following component failure play a dominant role. Nevertheless, Fig. 4.6 shows that there is a strong correlation between the height the debris obtains during its initial motion and the subsequent distance traveled. Kordi and Kopp (2011) also found that debris reaching higher heights also tends to travel at higher speeds. Numerical models of Eq. (4.5) utilizing uniform, smooth flow, with the uniform wind speed being the 3-s gust speed at the roof height, tend to capture the longest distances such objects can travel, but are unable to capture the variation without additional information.

### 4.2.5 Overall Flight Distances and Trajectories—Compact Debris

As mentioned, for compact debris it is usual to assume that the lift and moment coefficients are negligible so that the only force coefficient is the drag. Thus, the most common trajectories for gravel are ones where the debris is falling continuously. Trajectories for spheres (gravel) have been examined by Holmes (2004). More recently, Karimpour and Kaye (2012) and Moghim and Caracoglia (2014) examined the effects of turbulence and showed that it substantially increased the variation in the flight distances. Because there is no lift, compact debris requires a vertical wind speed component in order to rise. This occurs in the atmospheric surface layer of hurricanes, although the correlation between the vertical and horizontal wind components is negative, so that upward vertical gusts are usually associated with relatively quiet moments in the horizontal wind while the largest gust speeds are correlated with downward wind components. However, upward wind components occur in the separated flow and vortices caused by roof edges

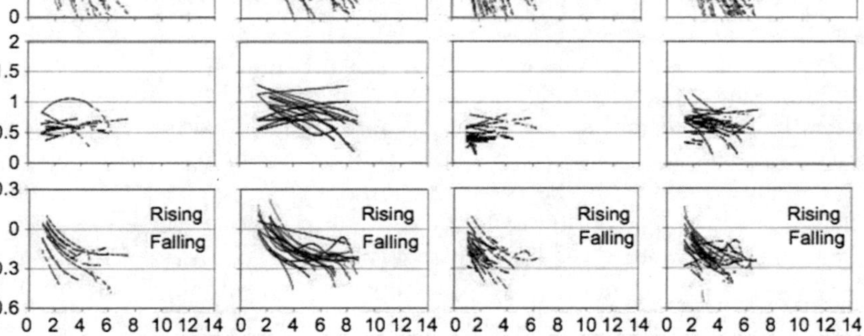


*Fig. 4.6. Trajectories and debris speeds for shingles originating from different locations on the roof of a 2-story house*

Source: From Kordi and Kopp (2011)

and in the flow field in the vicinity of the upper portions of the windward face of buildings, so the position and relative heights of buildings are important to these trajectories. Post-storm damage surveys have shown that glazing and windows at heights above the source building roof can be broken by gravel (or other broken glass), as discussed in Chapter 2, which is caused by these upward wind components.

## 4.3 FLIGHT SPEEDS

In uniform, smooth flow, it has been shown (Lin et al. 2007; Letchford et al. 2013) that the debris flight velocity follows the form

$$\overline{u} = 1 - e^{-\sqrt{C_1 K \overline{x}}} \tag{4.15}$$

where $C_1 \sim O(1)$ is a constant that depends on the type and shape of debris object.

Lin et al. (2007) and Letchford et al. (2013) provide values for compact and rod-like debris. For spheres, $C_1 = 1.0$, while for rods $C_1 = 1.6$ (Letchford et al. 2013). These curves show that the non-dimensional distance to the asymptotically limited wind speed is roughly $K\overline{x} \sim 4$ if $C_1 = 2$ and $K\overline{x} \sim 9$ if $C_1 = 1$.

For a rock of diameter 16 mm (0.63 in.) and weight of 5.4 g, $\overline{u} = 1 - e^{-\sqrt{0.022x}}$. For example, for a 7.3 m (24 ft) tall building and a basic wind speed [i.e., 3-s gust at 10 m (32.8 ft) in open terrain] of 193 km/h (120 mi/h), gravel may be expected to travel at least 20 m (65.6 ft). In this case, $\overline{u} = 0.48$, so that the dimensional horizontal speed would be 87 km/h (54 mi/h). Of course, one may expect substantial variation in the trajectory distance so that if it traveled 50 m (164 ft), the resulting speed would be 118 km/h (73 mi/h).

For a 2.4 m (8 ft) long 2 × 4 lumber with a weight of 4.1 kg (9 lb), the non-dimensional debris speed would be $\overline{u} = 1 - e^{-\sqrt{0.058x}}$. Fig. 4.7 depicts the solution to this equation as a function of horizontal displacement, indicating that 2 × 4 lumber traveling 20 m (65.6 ft) will have a speed of about 65% of the roof height gust speed, while at 40 m (131 ft) it will travel at about 80% of the gust speed. Of course, it must be emphasized that there would be a great deal of scatter in the landing locations, but flight distances of 20–40 m (65.6–131 ft) from 2-story houses would be expected (noting that there are no published studies considering these distances with sources from real buildings, possibly due to the extreme complexity in the initial conditions). The asymptotic limit of the speed for a 2 × 4 lumber is reached at $K\overline{x} \sim 5$, which translates to $x \sim 140$ m (459 ft). These speeds can be compared with standardized impact test requirements, as discussed in Chapter 7.

For plates, the form of Eq. (4.15) is not consistent with the asymptotic limits in Eq. (4.11). The data in Fig. 4.6 also show that the form $\overline{u} = f(x)$ as indicated by Eq. (4.15) does not tend to occur in debris flights originating from roofs of buildings. For plate debris with higher values of $K$ (i.e., lighter elements), such as plywood or oriented strand board roof sheathing panels or asphalt shingles, the

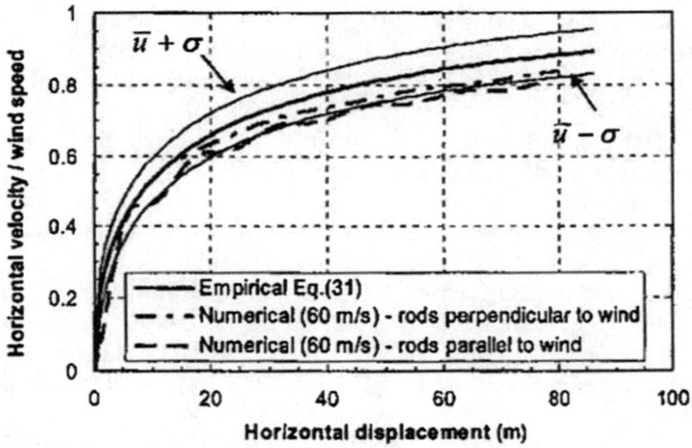


*Fig. 4.7. Horizontal speed of a 2.4 m (8 ft) long, 4.1 kg (9 lbs) 2 × 4 lumber as a function of horizontal displacement for a gust speed of 214 km/h (133 mi/h)*
Source: From Lin et al. (2007)

asymptotic limit provides a good estimate of the upper bound of the flight speed, using the 3-s gust speed at roof height. The range of speeds for plywood sheathing originating from a 2-story house is typically in the range of 20–90% of the 3-s gust speed at roof height (Kordi et al. 2010), while for shingles the range is 40–120% (Kordi and Kopp 2011). For roof tiles, which are much heavier, the non-dimensional flight distances are much shorter, typically, less than $K\overline{x} \sim 1$ (Kordi and Kopp 2011) for a 7.3 m (24 ft) high source building (e.g., a 2-story house). As a result, flight speeds are lower, typically, in the range of 30–60% of the 3-s gust speed at roof height, which is substantially lower than the asymptotic limit.

## *References*


ASCE. (2010). "Minimum design loads for buildings and other structures." *ASCE/SEI 7-10*, Reston, VA.

Baker, C. J. (2007). "The debris flight equations." *J. Wind Eng. Ind. Aerodyn.*, **95**(5), 329–353.

Holmes, J. D. (2004). "Trajectories of spheres in strong winds with application to wind-borne debris." *J. Wind Eng. Ind. Aerodyn.*, **92**(1), 9–22.

Holmes, J. D., Baker, C. J., and Tamura, Y. (2006). "Tachikawa number: A proposal." *J. Wind Eng. Ind. Aerodyn.*, **94**(1), 41–47.

Karimpour, A., and Kaye, N. (2012). "On the stochastic nature of compact debris flight." *J. Wind Eng. Ind. Aerodyn.*, **100**(1), 77–90.

Kind, R. J. (1974a). "Estimation of critical wind speeds for scouring of gravel or crushed stone on rooftops." *LTR-LA-142*, National Research Council of Canada, Ottawa, Canada.

Kind, R. J. (1974b). "Wind tunnel tests on some building models to measure wind speeds at which is gravel is blown off rooftops." *LTR- LA-146*, National Research Council of Canada, Ottawa, Canada.

Kordi, B., and Kopp, G. A. (2009). "Evaluation of the quasi-steady theory applied to windborne flat plates in uniform flow." *J. Eng. Mech.*, **135**(7), 657–668.

Kordi, B., and Kopp, G. A. (2011). "Effects of initial conditions on the flight of windborne plate debris." *J. Wind Eng. Ind. Aerodyn.*, **99**(5), 601–614.

Kordi, B., Traczuk, G., and Kopp, G. A. (2010). "Effects of wind direction on the flight trajectories of roof sheathing panels under high winds." *Wind Struct.*, **13**(2), 145–167.

Letchford, C., Lin, N., and Holmes, J. (2013). "Windborne debris in horizontal winds and applications to impact testing." Chapter 8, *Advanced structural wind engineering*, Y. Tamura and A. Kareem, eds., Springer, Tokyo, 197–215.

Lin, N., Holmes, J. D., and Letchford, C. W. (2007). "Trajectories of wind-borne debris in horizontal winds and applications to impact testing." *J. Struct. Eng.*, **133**(2), 274–282.

Lin, N., Letchford, C., and Holmes, J. D. (2006). "Investigation of plate-type windborne debris. Part I. Experiments in wind tunnel and full scale." *J. Wind Eng. Ind. Aerodyn.*, **94**(2), 51–76.

Masters, F., and Gurley, K. (2008). "Performance of embedded gravel roof systems in extreme wind loading." Univ. of Florida, Gainesville, FL.

Moghim, F., and Caracoglia, L. (2014). "Effect of computer-generated turbulent wind field on trajectory of compact debris: A probabilistic analysis approach." *Eng. Struct.*, **59**, 195–209.

Tachikawa, M. (1983). "Trajectories of flat plates in uniform flow with application to wind-generated missiles." *J. Wind Eng. Ind. Aerodyn.*, **14**(1–3), 443–453.

Tachikawa, M. (1988). "A method for estimating the distribution range of trajectories of wind-borne missiles." *J. Wind Eng. Ind. Aerodyn.*, **29**(1–3), 175–184.

Visscher, B. T., and Kopp, G. A. (2007). "Trajectories of roof sheathing panels under high winds." *J. Wind Eng. Ind. Aerodyn.*, **95**(8), 697–713.

Wills, J. A. B., Lee, B. E., and Wyatt, T. A. (2002). "A model of wind-borne debris damage." *J. Wind Eng. Ind. Aerodyn.*, **90**(4–5), 555–565.

# CHAPTER 5

# Damage from Impact (Experimental and Theoretical)

David Henderson*
Daniel Smith*
Russell Carter†
Douglas Smith†

Impact from wind-borne debris and the consequences of a breach in the building envelope during windstorms are design load criteria that need to be considered in addition to other design actions such as the wind pressure and dynamic load cycles. Debris mainly impacts windward walls (including doors and windows) and the upwind slope of roofs (Fig. 5.1), as discussed in Chapter 2. Damage investigations have shown that building envelopes constructed from light-weight cladding, glass windows, and roof tiles are especially susceptible to debris impact damage. Such damage can create openings in the building envelope, resulting in large internal pressures and an increased probability of more serious damage to the structure due to pressurization or water ingress.

## 5.1 INTRODUCTION

Observations after past windstorm events have shown impact damage to buildings caused by structural members such as timber (Fig. 5.2) or steel rafters, battens, and studs, and also by sections of roof that dislodged from adjacent buildings (Fig. 5.3). Often, a domino effect was evident, where debris from a building impacts another building downwind, which in turn increases internal pressure, causing the structure to fail and generate more debris that becomes available to impact other

*Cyclone Testing Station, James Cook University, Townsville, QLD, Australia.
†Department of Civil, Environmental, and Construction Engineering, Texas Tech University, Lubbock, TX, USA.

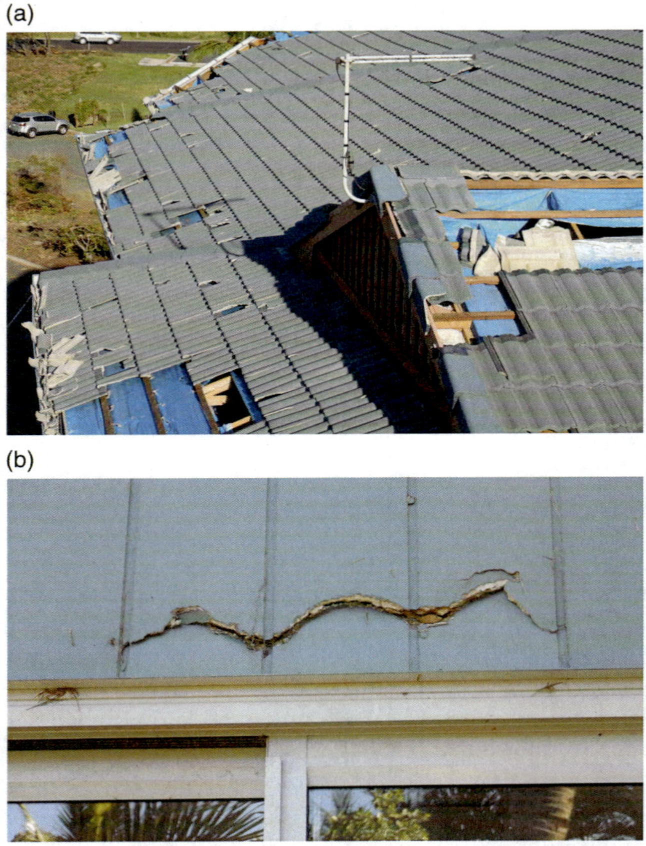

*Fig. 5.1. Damage to roofing tile system (left) and nearby wall cladding (right) from wind-borne roofing tile impact*

buildings downwind (Boughton et al. 2011). These observations highlight the importance of testing building envelope components for their resistance to wind-borne debris impacts. Minor (1994) highlighted the dramatic increase in losses due to a breach in the building envelope.

Predicting damage caused by wind-borne debris is challenging due to the characteristics of the debris, the properties of the impacted system, and the complex nature of the impact dynamics. Various techniques are used to analyze the effects of two objects colliding. A general technique is to consider the force of the impact as a quasi-static force. Techniques that are more detailed involve classic dynamics using kinetic energy methods for impact and momentum for both particles and rigid bodies. Impact dynamic theories combine dynamics and material behavior during impact, resulting in an understanding of how the objects behave

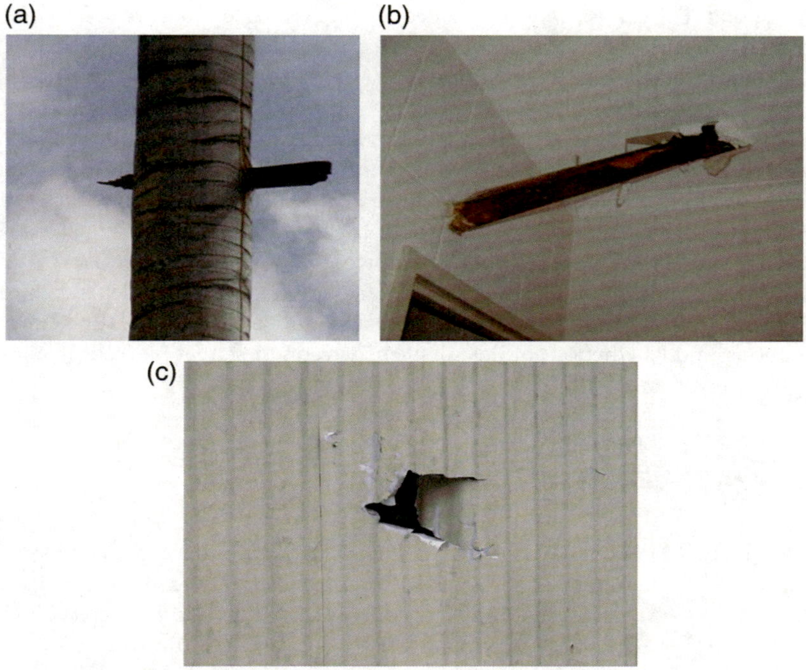


*Fig. 5.2. Rod-type wind-borne debris objects impacting trees (left), roofing and internal ceiling structure (center), and metal wall cladding (right)*

*Fig. 5.3. Complete roofing structure fails during a severe cyclone and becomes airborne and impacts nearby residential structures*

and respond. For clarity, wind-borne debris will be referred to as missiles and the impacted wall or roof system will be referred to as the target.

When the missile and the target collide, both objects begin to deform. The deformation does not occur instantaneously throughout the material. The deformation travels as a stress wave through the material at the speed of sound within the material. The stress wave propagates from the point of impact through both the missile and the target. Waves are affected by reflection, refraction, dispersion, and interference (Macaulay 1987) as they travel through the materials. As the target or missile (or both) deforms during impact the material(s) properties change, altering the wave speed. The progression of impact effects depends on the stress-strain properties of both the missile and the target materials the support conditions, and whether or not the missile penetrates the target (i.e. rate of loading and transfer of energy).

## 5.2 IMPACT VELOCITY

Wills et al. (2002) and Holmes (2007) demonstrated that the wind speed required to initiate flight of an object, $U_c$, is related to the density of the object and its size (see Chapter 3). An object that is dislodged from its fixings at a wind speed higher than $U_c$ is able to attain a maximum velocity approaching the wind speed (see Chapter 4). Larger objects with higher density require a higher wind speed than lighter, smaller objects to attain flight.

In addition to unsecured objects becoming missiles, debris from structural damage could be expected in windstorms when the gust wind speed exceeds about 125 km/h (78 mi/h). Under such conditions, many types of potential missiles can reach speeds of 53–106 km/h (33–66 mi/h) from a resting position within a short space of time and distance of travel. Reitano (2003) and Ginger et al. (2004) calculated time taken and distance traveled for a range of selected objects to reach such speeds by considering the flight patterns of compact, plate, and rod objects and applying aerodynamic parameters, as suggested by Wills et al. (1998). See Chapter 4 for more details.

Just as important as the impact velocity, the duration of the impact is a critical part of the puzzle in the resultant damage to the target from the missile. The greater the deformation of the target (and missile), the longer the impact duration leading to energy transfer and load distribution. The duration of impact for wind-driven debris is between 0.5 and 1.5 ms (Stephenson 1977).

## 5.3 KINETIC ENERGY AND MOMENTUM

The potential for damage will depend on the missile impact load and the resistance provided by the building envelope. Missile impact loading can be analyzed in terms of the missile impact velocity, $V_I$, and mass, $m$, of the missile in terms of kinetic energy $=0.5mV_I^2$ and its momentum $= mV_I$ along with the missile impact area and nose shape. Chen and Hao (2014) concluded that impact resistance should be determined using both kinetic energy and momentum analysis. Both the projectile mass and impact velocity play important roles in the resistance capacity.

Kinetic energy analysis provides one approach to describing impact and behavior. An example of an energy-based method used in material testing is the Charpy impact test. Carter (1998) summarizes that kinetic energy–based analysis is somewhat applicable for limited conditions such as when the target is made of a single solid material, and the missile and target behave in an elastic manner rebounding from one another at the end of the impact.

Conversely, impulse and momentum analysis is better suited to when the missile penetrates the target or when the target is a composite assembly. The impulse force changes the momentum of the missile and is a function of time. A short impulse creates high forces on the missile and target. The longer the impulse time on the material the lesser the resultant force magnitude. Longer time intervals may occur through deformation of the target or missile.

The occurrence of projectile deformation reduces the damage of the target compared with a similar geometry and mass. Masters et al. (2010) found that debris made of asphalt shingles required significantly higher momentum to damage a glass specimen than rod objects due to the rods' minimal deformation on impact compared with the highly deformable shingles impacting at the same momentum.

## 5.4 DEBRIS ORIENTATION ON IMPACT

Rigid bodies that are wind driven will have motion properties in all three dimensions. The motion may be described as the pitch, yaw, and rotation of the rigid body about the three-dimensional axis centered at the center of mass of the rigid body. The analysis of the rigid body behavior also depends on the location of the center of mass of the rigid body. The greatest force transfer from the debris to the target impacted system is when the debris has no pitch, yaw, or rotation and the center of mass is aligned with the axis of travel for the debris.

The majority of studies involve the normal impact on flat plate-like targets. Few studies involve oblique impact, projectile yaw, or deformable projectiles. Corbett et al. (1996) noted that obliquity does not significantly affect the penetration or rupture of a metal plate provided the angle of obliquity is less than 30°. However, the authors qualify this, noting that the increase in angle may lower the damage threshold because the presence of the projectile's yaw may change the failure mode of the target.

The worst case scenario for a rod-type missile impacting a target is in the end-on orientation because all of the missile's momentum and kinetic energy are concentrated in and transferred to the (comparatively) small cross-sectional area. For compact objects such as spheres the orientation of impact does not have a great effect because, by definition, this type of debris has similar dimensions in all three planes. For sheet objects, similar to the rod-type analysis, an end-on impact (i.e., the thickness of the sheet perpendicular to the target) is most critical because all of the sheet's momentum and kinetic energy would be transferred through the thin dimension, resulting in a shearing mechanism similar to a cutting action (chopping) (aside from the sheet-like object impacting face on and potentially increasing the surface area of the target, leading to an increase in wind load on the target (e.g., like roof cladding wrapping across power poles and lines).

Current test methods launch debris perpendicular (normal) to the surface of the target. Debris testing at Texas Tech University, TTU, indicates that the critical angle is 11°–13° from perpendicular (Carter and Stuckley 2001).

## 5.5 DEBRIS MATERIAL EFFECTS

The type of failure that occurs in metals is primarily dependent on the projectile nose shape, the ratio between the target thickness and the projectile diameter, and the material properties of the target material. In general, relatively thin, ductile targets when struck by blunt projectiles fail through discing, which is a thinning of the plate material under the impact zone leading to tensile tearing. This process is usually accompanied by substantial amounts of global discing of the target. Thick, brittle targets tend to fail by plugging a shear-dominated process in which a plug of material is separated from the surrounding target. Sharp-nosed projectiles cause failure by petaling in thin, ductile targets and plugging in thick, brittle targets. In all cases the amount of global deformation experienced by the targets is greatest at the ballistic limit, defined as just prior to breach in target, and decreases with increasing impact velocity.

## 5.6 STRUCTURE RESPONSE

The response of a component of the building envelope when impacted by wind-borne debris depends on the parameters related to the missile itself and the properties of the envelope component or system. The relevant parameters of the missile are its size, mass, shape, impact velocity, and orientation upon impact. The properties of the envelope are its material properties such as elastic modulus (ductile materials) or rupture modulus (brittle materials), the section properties determining the stiffness, the support conditions (i.e., distance of free span), and the location of impact.

The properties of the building envelope and structural support depend on the elastic and plastic response of the cladding, which in turn is dependent on the impact location. If the impact is closer to midspan the cladding may be able to deflect and undergo plastic deformation, thus, absorbing energy. In this impact region, Frye et al. (2012) measured reaction forces that related closely to a given missile's kinetic energy, independent of the missile's mass or impact velocity. Conversely, for impacts close to the cladding supports (e.g., girts, purlins, and studs) the cladding is restricted from deforming due to the proximity and stiffness of the supports. Frye et al. (2012) measured reaction forces closely related to the impacting missile's momentum, regardless of its mass and impact velocity (Fig. 5.4).

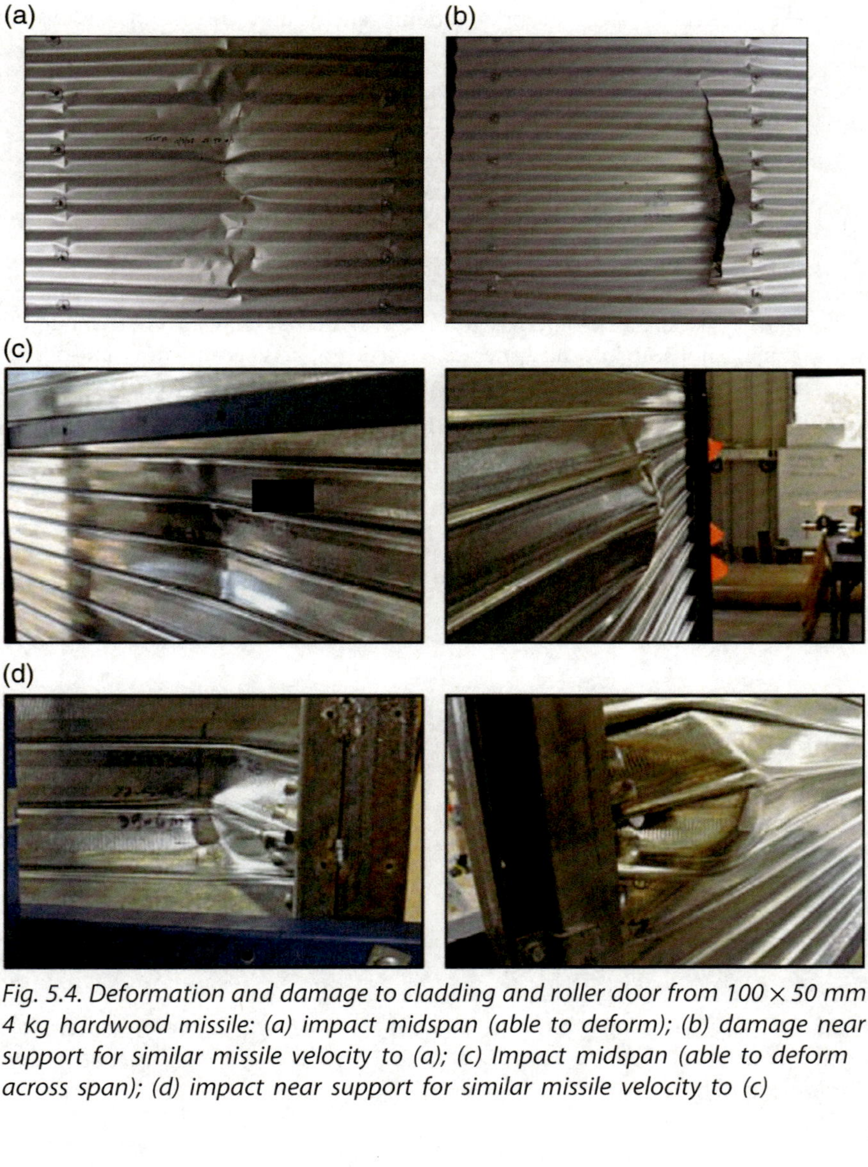


*Fig. 5.4. Deformation and damage to cladding and roller door from 100 × 50 mm 4 kg hardwood missile: (a) impact midspan (able to deform); (b) damage near support for similar missile velocity to (a); (c) Impact midspan (able to deform across span); (d) impact near support for similar missile velocity to (c)*

Structural systems are typically comprised of different materials and support conditions. Even glazing, which would appear to be a simple one element and material system, has a response that can be dependent on the glass fixity into the frame (i.e., the material and geometrical properties of the glass, sealant, and frame). The impact behavior of composite systems is thus complex and dependent on many variables. The behavior and failure are usually the result of a mixture of modes with one dominant mode. Zukas et al. (1982) defined the modes of failure and the contributing conditions as follows:

1. Brittle failure, typically called front-face spall in masonry and concrete. It is produced by local material inhomogeneity and anisotropy and large deformations.
2. Fracture results from the initial stress wave exceeding the material's ultimate strength (common in masonry structures).
3. Fragmentation, also called back-face scabbing, is the result of the compressive wave reflecting off the back of the target as a tensile wave. Fragmentation is common in materials that have a compressive strength greater than their tensile strength.
4. Ductile failure occurs when a malleable missile has a very high impact velocity leading the materials to behave hydrodynamically. This mode of failure is not seen in wind-generated missile impacts because the wind-borne missiles do not achieve the velocity required.
5. Plugging is a common failure mode for plywood and other types of sheathing material.
6. Petaling occurs when blunt-ended missiles impact the target at a velocity near the missile perforation threshold and is created by high torsional stresses after the initial stress wave has passed. Petaling involves plastic flow creating permanent flexure of the target (Fig. 5.5).

(a) (b)

*Fig. 5.5. Typical "petaling" (left) and "plugging" (right) type failure modes during experimental study (Laboy-Rodriguez et al. 2013) of roofing tile impacts on metal window shutters*

Suarez (1971) classified local response of the target material to missile impact as:

1. Elastic impact, in which the missile rebounds without damaging the target.
2. Elastoplastic impact, in which the missile rebounds after the target has started to permanently deform or deflect.
3. Plastic impact, in which the missile becomes embedded in the target.
4. Infra-plastic impact, in which the target is perforated by the missile.

This classification system is the starting point for applying appropriate analytical models or developing new models based on concepts from impact dynamics. This system can be expanded to include a more global consideration for structural response to the impact load using the following system described by Dallard (1984):

1. Elastic structural response (similar to billiard ball collisions).
2. Local plastic structural response in which the impact area exhibits plastic deformation but the structure responds elastically.
3. Global collapse in which a stable structure, such as a building, develops plastic hinges and fails.
4. Global collapse of an unstable structure that is not attached to the ground (such as an automobile).
5. Brittle structural response (primarily concrete structures) in which the compressive strength is greater than the tensile strength.

The classification of response allows the engineer to make some assumptions about the parameters of the missile impact. The assumptions include the amount of energy absorbed during the impact, the load deflection characteristics of the target, the peak force of the missile impact on the target, and the time of duration for the missile impact (Carter 1998). These response classifications are based on the configuration and response of the target. This is a response classification and does not account for the types of impacts or the failure mode of the material of the target. Analysis models exist for every classification listed above. The problem is that no single equation is applicable for every target or material that is represented by the same characteristics.

## 5.7 DEBRIS IMPACT TESTING

Test methods to determine performance of building components and systems, such as cladding, windows, and doors, to wind-borne debris impact have been developed in various countries. For example, ASTM E1886 (ASTM 2013) details methods to test complete glazing systems including frames, while ISO 16932 provides criteria and methodology for conducting tests on glazing allowing comparison across different glass products (excluding fenestration details).

Typically, the goal for all standardized test methods for product resistance to wind-borne debris, as opposed to other impact standards such as human impact, is to provide a level of protection to both the building (property) and bystanders (people). The protection to the property is from limiting the formation of a windward surface opening that could result in a sudden dramatic increase in internal pressure of the building [see Chapter 2 and Ginger et al. (2004)], as well as limiting the damage to the interior building and contents from wind and rain. The protection to bystanders is through the obvious resistance of the test specimen in stopping the missile's penetration of the building envelope. Some standardized test methods also set criteria for the minimal development of glass shards from impacted glazing to limit these secondary projectiles from impacting people in the vicinity of the window.

However, results from standardized testing are a measure of comparison between products and should not be considered as a replication of an actual cyclonic wind event since the missiles and wind loads will be different from the test method. It is this point that building designers should be considering in assessing design criteria such as whether to incorporate a large internal pressure from a breach in the building envelope or the possibly cheaper design option of assuming no breach or opening in the envelope. It would seem counterproductive to reduce the robustness of the whole building due to a test based on a standardized piece of timber or ball when damage surveys have shown debris can be in a broad range of forms from single timbers to almost complete roof structures.

ASTM developed its early wind-borne debris tests from some of Australia's early testing methods. However, in contrast to the United States, wind-borne debris impact testing in Australia [i.e., AS/NZS1170.2 and the public cyclone shelter design guidelines; see Standards Australia (2011) and Queensland Government Department of Public Works (2006), respectively] is predominantly qualitative (e.g., pass/fail outcome, quantitative measurements are not made). The only exclusion is the case of debris screens that are intended to be mounted as glazing protection, which requires a maximum deflection less than the clear distance between screen and glass, and the screen deflection has to be measured while the test specimen is subjected to the impact loading.

Through qualitative testing, a building system or component's resistance to wind-borne debris impact is assessed without the applied forces or reactions being measured or part of the assessment process, limiting the ability to infer or predict the likely response of another similar item. In addition, straightforward analytical solutions are not available, so product testing is required for certification. However, qualitative assessment after impact testing can still provide valuable insight into a component's behavior, such as observation of the failure mechanism or the extent of damage and deformation. Furthermore, these qualitative tests allow the determination of the most critical impact locations for typical envelope components and systems.

Masters et al. (2010) conducted experiments (Fig. 5.6) suggesting vulnerability to residential window glass by extracting the momentum threshold required to damage window glazing when impacted by roof shingles. Vulnerability curves were developed for unprotected window glass as a function of momentum, debris

*Fig. 5.6. Dowel launcher and glazing mounts for air cannon and drop test apparatus*
Source: Masters et al. (2010)

type, flight mode, and angle of impact. Experimental investigation was done for the momentum threshold required to damage typical annealed glass 3.2 mm (0.125 in) thick when impacted by shingles and wooden dowels. Experimental variables included shingle size (2), flight mode (spinning and tumbling), impact angle (45° and 90°), age of shingles (new and naturally aged), and impact speed.

Naturally aged, full-size shingles required a higher mean momentum than new full-size shingles to cause damage. It was also observed that glass has a lower damage momentum threshold for shingle debris in a tumbling flight mode than a spinning flight mode. This is because more deformation of the spinning shingle provides less elastic collision and less momentum transfer to the glass; 45° impact corresponds to lower damage than 90° impact. Masters et al. (2010) also found that shingles develop sufficient momentum to break window glass in Category 2 hurricanes on the Saffir-Simpson scale. A gust of 180 km/h (112 mi/h) is required to achieve a momentum with greater than 70% probability of damaging double strength–annealed residential window glass.

Masters et al. (2010) determined for that natural debris–size dowels the Category 1 hurricane local wind field is capable of imparting a momentum of 3 kg·m/s (0.67 lb·s) corresponding to an over 80% probability of failure for 90° impact. They found that shingle debris requires significantly higher momentum to damage a glass specimen in comparison to dowels. This is because the dowels undergo almost no deformation upon impact and thus impart more energy to the glass specimen compared with highly deformable shingles impacting at the same momentum. For both types of missile debris (shingles and dowels), debris weighing between 200 and 400 g (7 and 14 oz) would very likely achieve the necessary momentum to break building envelope components such as window glass if the local wind field is of low Category 2 intensity on the Saffir-Simpson hurricane scale, i.e., 161 km/h (100 mi/h). However, Category 1 wind speeds are also capable of imparting momentum with significant probability of breakage.

Some test methods and standards do require specific impact locations across the target (building product or system under evaluation). For example, ASTM E1886 requires debris impact locations in the middle of the glazing element adjacent to the edge and near the corner. Similarly, AS/NZS 4505 requires impact locations at midspan and adjacent to the panel edge (wind lock guides) for garage doors. The various test locations are to account for the differences in the target deformation and response, which can result in potentially different failure mechanisms.

## 5.8 APPLICATION

Sources of wind-borne debris are as varied as the items that make the natural and built environments. However, codified testing to evaluate building envelope performance from wind-borne debris uses impacts from rods or spheres representing rafters and gravel, respectively. Mitigation measures for wind-borne debris is a combination of minimizing sources of potential debris (e.g., tree pruning, neighborhood cleanup, or removal of rooftop gravel) and designing the building with robustness to account for an impact and breach of the building envelope. For example, the building design should incorporate potential for an increased load on the structure from internal pressure due to an opening in a windward face, that is, the building should not fail due to breach of one element.

### *References*

ASTM. (2013). "Standard test method for performance of exterior windows, curtain walls, doors, and impact protective systems impacted by missile(s) and exposed to cyclic pressure differentials." *E1886-13a*, West Conshohocken, PA.

Boughton, G. N., et al. (2011). "Tropical Cyclone Yasi structural damage to buildings." *CTS Tech. Rep. No. 57*, James Cook Univ., Townsville, Australia.

Carter, R. R. (1998). "Wind-generated missile impact on composite wall systems." M.S. thesis, Texas Tech Univ., Lubbock, TX.

Carter, R., and Stuckley, A. (2001). *Critical impact angle for medium class debris*, http://www.depts.ttu.edu/nwi/Pubs/ReportsJournals/ReportsJournals/Critical_Impact_CARTER.pdf (Dec. 07, 2017).

Chen, W., and Hao, H. (2014). "Experimental and numerical study of composite lightweight structural insulated panel with expanded polystyrene core against windborne debris impacts." *Mater. Des.*, **60**, 409–423.

Corbett, G. G., Reid, S. R., and Johnson, W. (1996). "Impact loading of plates and shells by free-flying projectiles: A review." *Int. J. Impact Eng.*, **18**(2), 141–230.

Dallard, P. R. B., and Miles, J. C. (1984). "Design tools for impact engineers." *Proc., Int. Conf. on Structural Impact and Crashworthiness*, Vol. **2**, Elsevier, London, 369–382.

Frye, U., Ginger, J., and Leitch, C. (2012). "Response of metal cladding systems to windborne debris impact." *Rep. TR59*, James Cook Univ., Townsville, Australia.

Ginger, J., Leitch, C., Henderson, D., and Reitano, C. (2004). "Flying debris impact in windstorms." *Proc., Int. Conf. on Building Envelope Systems and Technologies*, Sydney, Australia, ICBEST, Pennant Hills, Australia.

Holmes, J. D. (2007). *Wind loading of structures*, CRC Press, Boca Raton, FL.

ISO. (2016). "Glass in building—destructive-windstorm-resistant security glazing—test and classification." *ISO 16932:2016*, Geneva.

Laboy-Rodriguez, S. T., Smith, D., Gurley, K. R., and Masters, F. J. (2013). "Roof tile frangibility and puncture of metal window shutters." *Wind Struct.*, **17**(2), 185–202.

Macauley, M., ed. (1987). *Introduction to impact engineering*, Springer, Dordrecht, Netherlands.

Masters, F. J., Gurley, K. R., Shah, N., and Fernandez, G. (2010). "The vulnerability of residential window glass to lightweight windborne debris." *Eng. Struct.*, **32**(4), 911–921.

Minor, J. E. (1994). "Windborne debris and the building envelope." *J. Wind Eng. Ind. Aerodyn.*, **53**(1), 207–227.

Queensland Government Department of Public Works. (2006). "Design guidelines for Queensland public cyclone shelters." Brisbane, Australia.

Reitano, C. (2003). "Flying debris damage potential in windstorms." B.E. thesis, James Cook Univ., Townsville, Australia.

Standards Australia. (2011). "Structural design actions. 2: Wind actions." *AS/NZS1170. 2:2011*, Sydney, Australia.

Standards Australia. (2012). "Garage doors and other large access doors." *AS/NZS 4505:2012*, Sydney, Australia.

Stephenson, A. E. (1977). "Full-scale tornado missile impact tests." *Final Rep. to Electric Power Research Institute, Project No. 399*, Sandia Laboratories, Tonopah, NV.

Suarez, M. A. (1971). "Impactive dynamic analysis." *Proc., ASCE Structural Engineering Conf.*, ASCE, Reston, VA.

Wills, J. A. B., Lee, B. E., and Wyatt, T. A. (2002). "A model of wind-borne debris damage." *J. Wind Eng. Ind. Aerodyn.*, **90**(4–5), 555–565.

Wills, J. A. B., Wyatt, T. A., and Lee, B. E. (1998). "Warnings of high winds in densely populated areas." *UK IDNDR Flagship Programme, Rep. No. 4*, Thomas Telford, London.

Zukas, J. A., Nicholas, T., Swift, H. F., Greszczuk, L. B., and Curran, D. R. (1982). *Impact dynamics*, Wiley, New York.

# CHAPTER 6

# Wind-Borne Debris Impact Risk Modeling

J. Michael Grayson*
Weichiang Pang†

This chapter discusses the failure risk of the building envelope to wind-borne debris impacts in high-wind conditions. In order to assess wind-borne debris impact risk during windstorms, it is important to understand the basic principles of debris flight and the randomness associated with debris trajectories. This chapter provides a brief history of wind-borne debris simulation studies, the minimum modules required for the computational modeling of wind-borne debris, and the current state-of-the-art wind-borne debris trajectory models currently utilized in wind-borne debris risk and loss assessments.

## 6.1 WIND-BORNE DEBRIS SIMULATION

### 6.1.1 Early Wind-Borne Debris Trajectory Research

The earliest wind-borne debris trajectory models were born from the necessity to evaluate the risk of wind-borne debris produced by tornados affecting critical facilities, not necessarily to prevent damage to the facility, but to make sure that the facility maintains integrity to protect the public (Dunlap and Widener 1971). These early risk models were simple in nature and ultimately reduced the projectiles down to a point that only considered elementary forces acting on the debris (e.g., Lee 1974; McDonald et al. 1974). Simiu and Cordes (1976) treated wind-borne debris as a point [e.g., a three degrees-of-freedom (DOF) trajectory model] with only a drag force acting on it, but developed the concept of the average drag coefficient to account for the random tumbling of wind-borne debris during flight.

Redmann et al. (1976, 1978) developed a six DOF wind-borne debris trajectory model to account fully for the drag, lift, and side forces applied to the

*Department of Civil Engineering, The Citadel, Charleston, SC, USA.

†Glenn Department of Civil Engineering, Clemson University, Clemson, SC, USA.

debris while also tracking the pitch, yaw, and roll of the debris. This model incorporated a realistic near-ground tornado wind field, experimentally derived aerodynamic coefficients for the debris shapes of interest, theoretical damping coefficients for the debris, the capability to build up debris shapes from elementary parts, and an adaptive integration and time-stepping program to maximize the computational efficiency of the trajectory model. Twisdale et al. (1979) expanded upon the concept of the six DOF wind-borne debris trajectory model and introduced a random orientation six DOF model, which included the drag, lift, and side forces acting on the wind-borne debris during flight. As the name implies, the orientation of the debris during flight was chosen randomly to incorporate the stochastic nature of the debris during flight with the actual debris trajectory being a deterministic process.

### 6.1.2 The Experiments of Tachikawa

Tachikawa (1983) utilized Newton's equations of motion to describe the motion of generic plate debris in a uniform wind field. As discussed in Chapter 4, this research ultimately provided assumptions for integrating Newton's equations of motion, and identified the modes of motion of a flat plate during flight. Arguably one of the most significant contributions made by Tachikawa was the development of a non-dimensional parameter that consisted of the ratio of aerodynamic forces to gravitational forces. Little did Tachikawa know it at the time, but this parameter, designated "*K*" by Tachikawa, was the controlling parameter in determining wind-borne debris trajectories for all debris types (Holmes et al. 2006a). Tachikawa (1988) further proposed a method for estimating the distribution range of wind-borne debris trajectories. Wind tunnel tests were used to identify the associated distribution characteristics required to estimate the spatial distribution of the wind-borne debris trajectories, and to further investigate which factors influence these characteristics (e.g., wind velocity profile, ascending flow, vertical reaction force, and the scale of the wind-borne debris). Tachikawa (1988) concluded that the distribution parameters of the range of wind-borne debris trajectories could be estimated from the probability distribution of the debris lift coefficient; however, Tachikawa stated that further studies were needed to ensure the validity of this method.

### 6.1.3 The Evolution of Wind-Borne Debris Trajectory Research

Twisdale et al. (1996) highlighted the interdependence of wind-borne debris trajectories and impact. This led to the development of the hurricane missile (HURMIS) risk analysis methodology, which was not only capable of providing information on wind-borne debris trajectories and speeds, but also wind-borne debris impact parameters. Twisdale et al. (1996) provided a comparison that illustrated the benefits and costs of previous wind-borne debris trajectory models before finally settling on the random orientation six DOF model developed by Twisdale et al. (1979). The random orientation six DOF wind-borne debris trajectory model utilized in the HURMIS risk analysis methodology considers the drag, lift, and side forces acting on the wind-borne debris during flight while using a periodic reorienting of the debris body to simulate the tumbling experienced by the debris due to the applied forces. This method of reorientation allowed for better wind-borne debris trajectory prediction estimates over particle trajectory models with only a minimal decrease in computational efficiency.

Wills et al. (2002) sought to calculate the damage potential of wind-borne debris by relating the aerodynamic forces experienced by wind-borne debris with the damage that it inflicted upon impact with a structure. This required that Wills et al. (2002) define all forces acting on the debris before release (i.e., the fixity of the debris) as well as the aerodynamic forces during flight, as discussed in Chapter 3.

Holmes (2004) studied the trajectories of compact debris by theoretically and analytically modeling the trajectories of spheres in extreme wind events. This research was instrumental in illustrating that the vertical air resistance experienced by compact debris during flight significantly increases the debris flight time and the horizontal displacement and velocity, while decreasing the vertical velocity. It also illustrated that the inclusion of atmospheric turbulence in wind-borne debris trajectory calculations has little influence on the mean trajectory results, but readily affects the variability of individual compact debris trajectories, with the largest variability seen in the individual horizontal displacements and velocities.

Lin et al. (2006) provided a comprehensive assessment that utilizes wind tunnel and full-scale experiments, and Holmes et al. (2006b) analytically computed wind-borne debris trajectories to quantify the aerodynamic flight characteristics of plate-type debris. The wind tunnel and full-scale tests performed by Lin et al. (2006) identify the Tachikawa number ($K$) as the predominate determinant of wind-borne debris horizontal trajectories, and further led to the development of empirical equations that can be used to estimate the horizontal distance and velocity of plate-type debris. Holmes et al. (2006b) provided the analytical calculations to compare with the experimental tests performed by Lin et al. (2006) and generally provide good agreement with the experimental results while also confirming the findings of Lin et al. (2006) that the Tachikawa number is the governing determinant of wind-borne debris trajectories.

Lin et al. (2007) compared experimental and numerical solutions to the trajectories of compact and rod-type debris. This study supplemented the work of Lin et al. (2006), since it presented empirical equations that can be used to estimate the horizontal distance and velocity of compact and rod-type debris. Lin et al. (2007) further concluded that reporting an average ratio of debris speed to wind speed, over the duration of a trajectory as done by Wills et al. (2002), does not provide useful information since the wind-borne debris tends to accelerate toward the wind speed and asymptotically approaches a ratio of 1 as a function of horizontal displacement. Lin et al. (2007) also noted the importance of including the probabilistic nature of wind-borne debris in wind-borne debris risk models. Little attention had been given to the probabilistic nature of debris trajectories up until that point since it was necessary to study the debris trajectories deterministically to ensure that the parameters influencing wind-borne debris displacements and speeds were identified in order to appropriately address the trajectories in a probabilistic manner.

Baker (2007) studied the two-dimensional motion of compact and sheet debris utilizing alternative non-dimensional forms of Newton's equations of motions from that of Tachikawa (1983). Baker concluded that compact debris flight was controlled by a single dimensionless parameter, essentially the inverse of the Tachikawa number, within this dimensionless scheme, as discussed in Chapter 4. Sheet debris flight is much more complex since it has several different rotational flight modes (e.g., clockwise, anti-clockwise, and no rotation) and exhibits a sensitivity to boundary conditions in addition to having several controlling parameters that require values typical of chaotic systems. Rod debris trajectories were not considered within this research in two dimensions due to the complex three-dimensional rotations that are typically seen during rod debris flight. Baker (2007) further supplemented the findings of Holmes (2004) by noting that the consideration of atmospheric turbulence in the calculation of the compact, as well as sheet, debris trajectories did contribute a noticeable variation in the overall wind-borne debris trajectories.

Visscher and Kopp (2007) studied the flight trajectory of scaled roof sheathing panels in a wind tunnel captured by high-speed digital video. These tests identified that the sheet debris exhibited two different modes of flight when considering similar initial conditions, i.e., translational and autorotational, with the autorotational mode of flight exhibiting the greatest debris trajectory distance, speed, and variability. Tachikawa (1983) also noted the wide range of trajectories of rotating flat plates and attributed this variation in trajectories to the Magnus effect. Visscher and Kopp (2007) concluded that the presence of different debris trajectory flight modes originating from essentially the same nominal conditions suggests that atmospheric turbulence plays a significant role in the aerodynamic forces experienced by the debris, most likely during the initial release of the debris and movement into the wind field. This suggests that the debris trajectories are even more sensitive to specific flight conditions than previously suggested by Baker (2007).

Richards et al. (2008) developed a deterministic six DOF wind-borne debris trajectory model that provided the three-dimensional flight motion of rectangular plate- and rod-type debris. Concurrent wind tunnel tests on flat plates and long rods determined that the aerodynamic force coefficients of the test specimens were heavily dependent on both the angle of attack and the tilt angle. Richards et al. (2008) also observed that the mean drag force and moments acting on a plate during flight were modified by the rotation of the plate similar to that described by Tachikawa (1983) and Baker (2007). Richards et al. (2008) also noted that the circular plate dispersion calculated using the deterministic model was similar to that reported by Tachikawa (1988) in that the diameter of the containing circle increased with an increase in the plate side length aspect ratio. Results obtained from the presented deterministic wind-borne debris trajectory model showed that the horizontal speed of the wind-borne debris routinely exceeded 90% of the wind speed if given enough flight time, and could potentially travel faster than the wind speed if optimal conditions were met with vertical momentum being converted to horizontal momentum and autorotation of the plate.

Noda and Nagao (2010) also developed a deterministic six DOF wind-borne debris trajectory model, but they focused their study exclusively on the three-dimensional motion of plates. A range of angles that defined the horizontal and vertical angles of attack were tested within a wind tunnel to obtain the aerodynamic forces that occurred on the plates during flight. Calculated trajectories were used to determine if the Tachikawa number ($K$), the plate aspect ratio, or both had any effect on the debris trajectories. Results from the study concluded that an increase in the Tachikawa number increased the flight distance of the wind-borne debris proportionally. Additionally, the extent of the debris dispersion (i.e., the diameter of the debris scatter region) was affected not only by changes to the Tachikawa number, but also by the plate aspect ratio and the wind speed.

### 6.1.4 Transition to Probabilistic Wind-Borne Debris Trajectory Models

Initially, probabilistic wind-borne debris trajectory models relied on basic probability distributions to determine specific quantities of interest pertaining to wind-borne debris trajectories that occur when debris is released into the wind. Lin and Vanmarcke (2008, 2010) developed a probabilistic wind-borne debris trajectory model that proposes the use of a Gaussian distribution from the debris landing positions and the horizontal debris impact momentum. The assumption of the Gaussian (normal) distribution for the debris landing positions was obtained from previous research by Tachikawa (1983), which noted that debris impact locations were for the most part uniformly distributed within circles from the center of a catch net. Lin and Vanmarcke (2008) noted that the preferred stochastic model for estimating debris horizontal impact momentum is a lognormal distribution, but that the Gaussian distribution was chosen as a simplification of the analytical calculations. The complexity of physical wind-borne debris trajectories and a lack of historical data necessitated that the parameters for the probability distributions be estimated from empirical models developed by Lin et al. (2006, 2007). Lin and Vanmarcke (2008) then utilized Poisson random measure theory to incorporate the probabilistic wind-borne debris trajectory model into a wind-borne debris risk model.

Lin and Vanmarcke (2010) extended their previous work (i.e., Lin and Vanmarcke 2008) to make it more generalized and thus more applicable to vulnerability and reliability analyses. The aim of the stochastic debris flight trajectory model presented was to obtain (1) the probabilistic distribution of debris landing positions, and (2) the conditional probability of horizontal impact momentum, which is an extension of the horizontal debris velocity (Lin and Vanmarcke 2010). Due to limited full-scale storm damage information, the parameter estimation was obtained from a combination of empirical observations, numerical simulations, and engineering judgment.

Grayson et al. (2012) developed a mechanics-based probabilistic wind-borne debris trajectory model by combining deterministic wind-borne debris trajectory models, such as those of Richards et al. (2008) and Noda and Nagao (2010), with statistical information obtained from scale-model tests performed by Kordi et al.

(2010). Assuming a Gaussian distribution, the angle of attack and tilt angle of the wind-borne debris identified by Richards et al. (2008) were treated as random variables within the deterministic model, and Monte Carlo simulations were performed utilizing a test matrix of coefficient of variation values. The mean, standard deviation, and skewness of the debris trajectories for a full sheet of roof sheathing [i.e., 0.0127 m (0.042 ft) × 1.22 m (4 ft) × 2.44 m (8 ft) in dimension] at each of the values in the test matrix were compared to the equivalent full-scale debris landing statistics from Kordi et al. (2010) utilizing the square root of the sum of the squared error technique to develop surface plots of the error for both the angle of attack and tilt angle. The minimum error identified from the surface plots coupled with visual comparisons to the wind tunnel test data provided the coefficients of variation to be used within the probabilistic wind-borne debris trajectory model for each of the wind directions tested by Kordi et al. (2010). Linear interpolation of the coefficient of variation for the angle of attack and the tilt angle between the tested wind angles provided the full range of wind angle values (i.e., −90° to 90°) that could potentially be encountered during a hurricane wind event.

## 6.2 WIND-BORNE DEBRIS RISK MODELING

### 6.2.1 The Incorporation of Wind-Borne Debris in Risk Modeling

Holmes (2010) provided an assessment of wind-borne debris risk modeling that included (1) the essential elements of damage risk models, (2) results from wind-borne debris trajectory models that are relevant to damage risk models, (3) the merits of impact momentum versus kinetic energy, (4) the use of data from generic impact tests in damage risk models, and (5) the variation of debris impact damage with wind speed. Holmes (2010) concluded that damage from wind-borne debris impacts can be almost as costly, if not as costly, as damage attributed to direct wind loading. The hope is that wind-borne debris impact damage may be accounted for by a simple factor taken over the direct wind loading damage. However, it should be realized that debris impact damage is closely correlated to direct wind loading damage. Wind-borne debris damage is extremely difficult to quantify compared to damage caused by direct wind loading. This is attributed to the inherent randomness of debris flight generation, trajectory, and impact. In addition, debris released by upstream buildings may damage the walls and roofs of downstream buildings and set off a chain reaction effect, which results in additional debris released by those impacted buildings, as discussed in Chapter 2. Because wind losses are the combined effects of damage from both direct wind loads and debris impact, many loss assessment models for high-wind events are derived from empirical relationships observed from past windstorms, in particular, those proprietary in-house models used by insurance and reinsurance companies. With recent advancements in modeling techniques and the availability of high-performance computing, wind loss assessment models including those incorporating flying debris are shifting from an empirical approach to a more engineering-based approach.

In order to take an engineering-based modeling approach to wind-borne debris loss assessments, several key submodules must be incorporated to ensure that all aspects of the loss assessment are considered (Twisdale et al. 1996; Lin and Vanmarcke 2008; Holmes 2010). These key submodules must consist of, at a minimum, the following:

1. *Building inventory models* (see Fig. 6.1) are needed to establish the loss assessment simulation domain. The spatial distribution of buildings, as well as their dimensions and shapes along with the locations and orientations of vulnerable components (e.g., windows and doors), should be included in the building inventory model. The building inventory model must include the resistances of building envelopes (e.g., uplift capacities of roof sheathings and coverings) and the impact resistances of the vulnerable components. Recent studies (e.g., Grayson et al. 2013; Grayson 2014) have begun to incorporate more detailed building inventory models that incorporate more aspects of residential developments in order to facilitate the use of mechanics-based damage models.
2. *A wind field generation module* is needed to generate the time histories of the peak gust wind speed and wind direction experienced by the study domain (see Fig. 6.2). For hurricanes or tropical cyclones, a gradient wind field is first generated at the top of the boundary level. The mean gradient wind speed is then converted into surface wind speed, usually at a standard reference height of 10 m (33 ft) using an assumed boundary layer model (e.g., Vickery et al. 2009) or a simple gradient-to-surface wind conversion factor (e.g., Sparks and Huang 2001). For tornadoes, Redmann et al. (1976, 1978) derived azimuthal and radial velocity components of the near-ground wind field. However, advances in theory, modeling, and observation techniques (e.g., Kosiba et al. 2008) and physical and numerical tornado simulators (e.g., Haan et al. 2008; Mishra et al. 2008; Maruyama 2011) have provided opportunities to develop more accurate near-ground tornadic wind field models that can be utilized in wind-borne debris risk modeling.

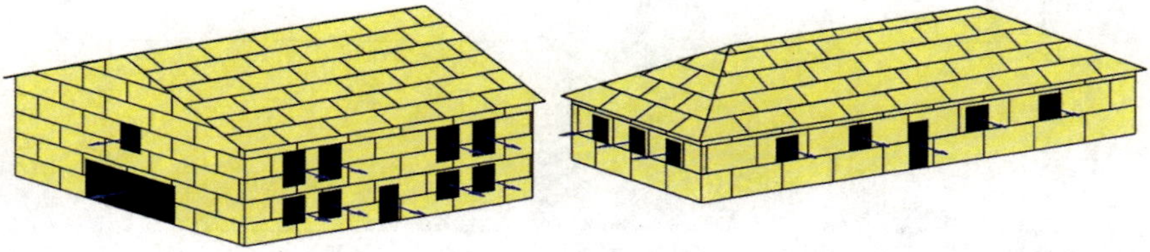

*Fig. 6.1. Example of building inventory models illustrating the exterior sheathing layout*
Source: Adapted from Grayson et al. (2013)

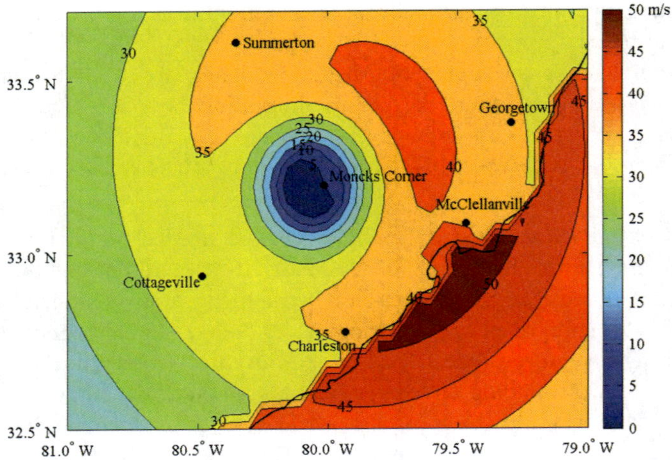


*Fig. 6.2. Surface wind speed contours for Hurricane Hugo after landfall near Charleston, SC, in 1989*
Source: Adapted from Grayson et al. (2013)

3. *A debris generation module* is needed to establish the types of debris, in addition to the amount and time evolution of debris released into the wind stream (see Fig. 6.3), encountered during a windstorm event. Wind-borne debris can come from the envelopes of low-rise buildings, trees, and to a lesser extent from miscellaneous items around the house (e.g., mailboxes or fences), as discussed in Chapters 2 and 3. The debris generation model must be able to convert wind speeds from the wind field model into wind pressures (i.e., demand). A building component is released into the wind

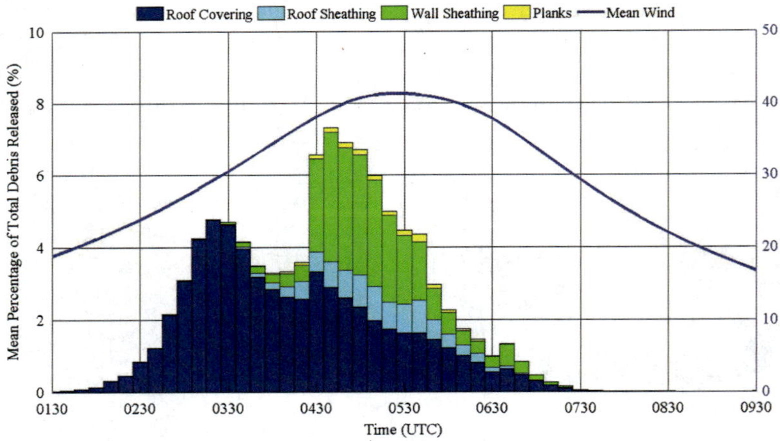


*Fig. 6.3. Time evolution of wind-borne debris release during a hurricane event*
Source: Adapted from Grayson et al. (2013)

stream as debris when the wind pressure or suction exerted on the particular building envelope component (e.g., roof covering or sheathing) exceeds the resistance defined in the building inventory model. In other words, the debris generation model can also be viewed as a wind failure model, which is the most basic component required in a conventional wind damage or loss assessment model.

4. *A wind-borne debris trajectory module* is needed to determine the impact locations and velocities of wind-borne debris on downstream structures. The wind-borne debris trajectory module must account for the inherent randomness of the debris trajectories (Lin and Vanmarcke 2008; Grayson et al. 2012). Ideally, a mechanics-based three-dimensional probabilistic debris flight trajectory module (e.g., Grayson et al. 2012) would be utilized to provide the most realistic approximation of the actual wind-borne debris trajectories (Fig. 6.4). However, two-dimensional debris flight models (e.g., Tachikawa 1988; Holmes 2004; Lin et al. 2007; Baker 2007) with incorporated randomness are useful in estimating horizontal impact locations and velocities. The debris trajectory module should also include, at a minimum, the weight (or mass) in order to calculate the impact kinetic energy or momentum depending on the quantity that is appropriate to assess any damage of the affected location.
5. *A wind-borne debris impact and damage module* is needed to detect impacts and to assess impact damage to affected structures. Presently, a majority of building components are rated on a pass/fail basis when subjected to wind-borne debris impact testing, as discussed in Chapters 5 and 7. Therefore, significant work is needed to identify and incorporate threshold impact damage levels to structural and building envelope components subjected to wind-borne debris. Impact damage modules in the future will most likely require the development of an extensive database that would be utilized to determine the extent of the damage due to wind-borne debris impacts (Fig. 6.5).

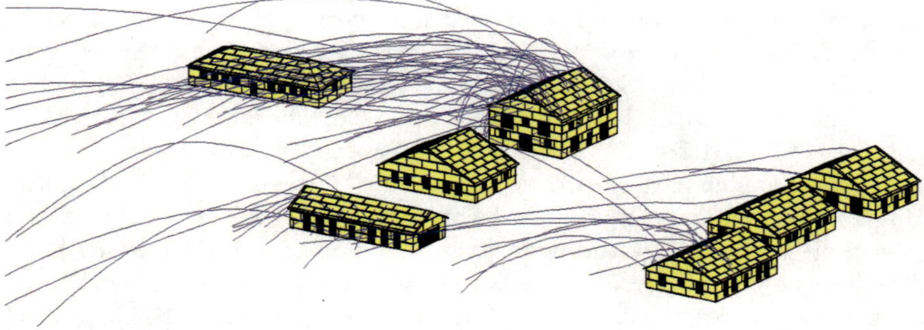

*Fig. 6.4. Probabilistic wind-borne debris trajectories generated from a detailed building inventory during the passage of a simulated Hurricane Hugo*
Source: Adapted from Grayson et al. (2013)

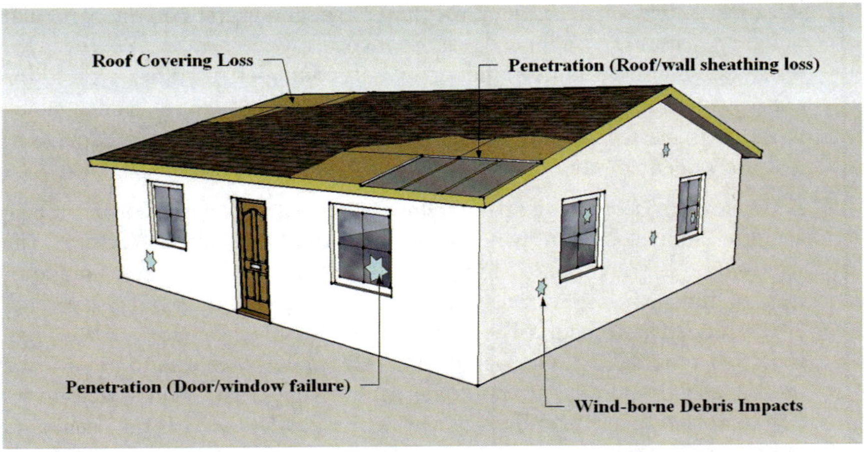


*Fig. 6.5. Generation of and damage due to wind-borne debris*
Source: Adapted from Grayson (2014)

### 6.2.2 Wind-Borne Debris Risk Modeling Using an Engineering-Based Approach

Until recently, few studies have dealt with wind-borne debris risk and loss assessment that were not, for the most part, empirical studies sanctioned by government agencies or insurance companies (Holmes 2010; Lin and Vanmarcke 2010). The earliest examples of risk assessment studies that focused on the damage potential of wind-borne debris were mainly initiated by the power industry and focused on tornado-generated wind-borne debris (e.g., Lee 1974; Simiu and Cordes 1976; Redmann et al. 1976, 1978; Twisdale et al. 1979). It was not until Twisdale et al. (1996) estimated the mean damage risk from wind-borne debris within a residential area subjected to a hurricane that the focus shifted toward low-rise residential developments. The Twisdale et al. (1996) report gave rise to several loss assessment models that incorporated wind-borne debris utilizing an engineering-based approach, for example, the Hazards-United States Multi-Hazard (HAZUS-MH) model (Vickery et al. 2006a, b; FEMA 2007) and the Florida Public Hurricane Loss Model (FPHLM) (Gurley et al. 2005; Pinelli et al. 2011).

The HAZUS-MH model uses an explicit damage and missile transport approach to estimate wind-borne debris parameters based on the Twisdale et al. (1996) model. However, the wind-borne debris model within the HAZUS-MH model is used only to generate impact energy and momentum risk curves and it does not explicitly consider wind-borne debris in the damage model (Vickery et al. 2006a). Wind-borne debris types considered in the HAZUS-MH model for the residential building model include roof tiles and shingles, roof sheathing panels, and structural members, in addition to whole roofs. In additon, the HAZUS-MH model employs a commercial wind-borne debris trajectory model that incorporates commercial building envelope damage from wind-borne gravel debris, which can be generated from aggregate-ballast roof systems.

The FPHLM (Gurley et al. 2005; Pinelli et al. 2011) utilized a debris risk model that was similar to the Twisdale et al. (1996) model. However, the FPHLM did not incorporate numerical simulations, but rather empirical observations from post-damage field assessments of residential areas, to calibrate the wind-borne debris parameters; therefore, the resulting wind-borne debris risk model was greatly simplified. Furthermore, the FPHLM only considers wind-borne debris impacts to windows from upstream debris sources, such as roof cover, and further makes the assumption that there is a high likelihood that an impacted window will be damaged (Pinelli et al. 2011).

Lin and Vanmarcke (2008) expanded upon the assumption of the Poisson distribution initiated by Twisdale et al. (1996) by illustrating that the total number of impacts that exceed an assumed damage threshold also follows a Poisson distribution. Lin and Vanmarcke (2008) isolated the study area to prevent the introduction of any outside influences, which is a typical assumption for the majority of current wind-borne debris risk assessment models in use today. Lin and Vanmarcke (2008) link together the stochastic processes associated with wind-borne debris (i.e., debris generation, flight, and impact) in order to aggregate the damage from each type of debris originating from each house within the study region, which in turn avoids the implied assumption of risk uniformity throughout the study region. Lin and Vanmarcke (2010) further defined the debris risk modeling problem as one of identifying (1) wind-borne debris generation, (2) debris flight trajectories, and (3) debris impact. Hurricanes produce debris that originates primarily from elevated sources (e.g., house roof coverings and roof sheathing), whereas tornados contain faster wind speed and updraft components that are capable of picking up and injecting debris located along the ground. Therefore, the principal quantities of interest in hurricane-induced wind-borne debris are (1) the horizontal debris velocity and (2) the debris displacement (Lin and Vanmarcke 2010).

A significant portion of losses experienced by residential structures during a hurricane wind hazard event are typically due to a breach in the building envelope of a structure. While it is difficult to determine the exact amount of damage that can be attributed to wind-borne debris impact, post-storm damage assessments have revealed that wind-borne debris impact can play a substantial role in these losses. In an effort to quantify these losses due to the breach of the building envelope from wind-borne debris impacts, many loss models began to focus heavily on the performance of the building envelope during extreme wind hazard events.

Lin et al. (2010) and Yau et al. (2011) applied the wind-borne debris risk assessment methodology presented by Lin and Vanmarcke (2008, 2010) to develop a structural vulnerability model that explicitly considered the relationship between wind-borne debris impact and the internal pressurization of a structure above and beyond that of the HAZUS-MH model or the FPHLM. One focus of these studies was to quantify the chain reaction effect that has been illustrated in post-hurricane damage surveys. Dao et al. (2012) further developed a model that investigated the vulnerable components of a building envelope by determining the probability of damage to windows within a residential housing development subjected to a hurricane wind event. The study by Dao et al. (2012) also illustrated the relationship between the time evolution of the wind-borne debris produced and the subsequent building envelope breach as the hurricane passes the development, which leads to considerable damage to building contents.

Grayson et al. (2013) developed a mechanics-based building envelope failure assessment (BEFA) framework that focused on smaller, more detailed residential developments. The BEFA framework exhibited several differences from preceding assessment models. The BEFA model explicitly modeled the homes within a residential development at the component level, from the roof covering and roof and wall sheathing to the actual fenestration size and layouts, as opposed to aggregating damage from a prototypical house. Additionally, the BEFA model explicitly tracked and incorporated wind-borne debris impact damage utilizing a mechanics-based three-dimensional probabilistic wind-borne debris trajectory module, as opposed to using damage curves based on statistical and/or empirical models. However, the detailed approach taken by Grayson et al. (2013) presented new challenges in determining how to quantify the amount of damage attributed to various types of component failure, in addition to damage attributable to wind-borne debris impact. The development of an analytical damage quantification method was necessary to combine the type and amount of damage incurred to the building envelopes of individual homes within a residential development due to direct wind loading and wind-borne debris impact (Grayson 2014; Grayson and Pang 2015). However, a limitation of the modeling approach by Grayson et al. (2013) is that the assessment is limited to small residential developments due to the computational intensity required by the detailed building inventory and wind-borne debris trajectory modules.

## 6.3 WIND-BORNE DEBRIS DAMAGE ASSESSMENT AT A RESIDENTIAL DEVELOPMENT LEVEL

Application of the BEFA model (Grayson 2014) provides an in-depth look at the synergistic relationship that occurs between homes within a residential development due to the production of wind-borne debris during a windstorm event. Grayson (2014) subjected a 38-home residential development near Moncks Corner, SC, to 10 probabilistic synthetic hurricane events that possess a peak 3-s gust surface wind speed representative of the 700-year mean recurrence interval design wind event for that particular location. The homes within the residential development were either un-retrofitted for the wind events or outfitted completely with robust upgrades to the roof sheathing attachment, roof covering, personal and garage doors, and the inclusion of impact protection systems on all windows. The residential development was modeled with five different levels of the percentage of homes retrofitted ranging from 0% up to 100%. The retrofitted homes for the 25%, 50%, and 75% cases were randomly selected for each simulation to remove the influence of the spatial layout of the homes from the assessment.

Figure 6.6 illustrates the time evolution of the mean building envelope survival of the residential development subjected to 10 synthetic hurricane events separately for the un-retrofitted and retrofitted homes for the five retrofit cases. Grayson (2014) presents the results of Fig. 6.6 as the complement of the mean building envelope damage of the residential development obtained from the BEFA model (i.e., the mean building envelope survival of the residential development) to facilitate future community resilience studies. The residential development receives very minor to no damage with 100% of the homes retrofitted with an increase in the damage to the retrofitted homes as the retrofit percentage decreases. The same trend is exhibited for the un-retrofitted homes although the damage experienced by the residential development with none of the homes retrofitted is approximately 50% greater than the residential development with all of the homes retrofitted. It is evident from Fig. 6.6 that the un-retrofitted homes are contributing to the damage of retrofitted homes in some manner because there is little to no damage to the residential development with 100% of the homes retrofitted. Figure 6.7 provides further insight into this phenomenon by inspecting the final mean building envelope survival (i.e., the survival values in Fig. 6.6 obtained at hour 32). The results from Fig. 6.7 examine the contributions of the failure types illustrated in Fig. 6.5 for the retrofitted and un-retrofitted cases.

Closer inspection of Fig. 6.7 shows the only quantity that is variable from case to case is the damage incurred to both the retrofitted and un-retrofitted homes due to wind-borne debris impact. The un-retrofitted homes in all cases sustain damage from building envelope penetrations due to sheathing loss, window and door failure, and roof covering loss, while the majority of the damage sustained by the

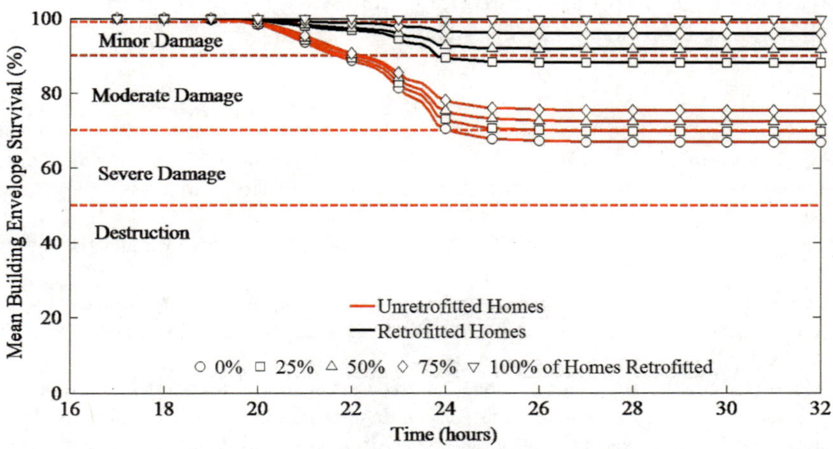


*Fig. 6.6. Time evolution of the mean building envelope survival of a residential development subjected to 10 synthetic hurricane events*
Source: Adapted from Grayson (2014)

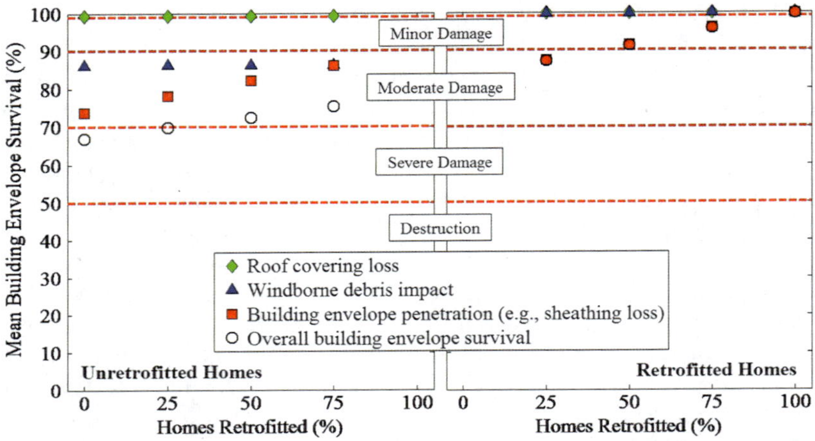


*Fig. 6.7. Final mean building envelope survival contributions of a residential development subjected to 10 synthetic hurricane events*
Source: Adapted from Grayson (2014)

retrofitted homes comes from wind-borne debris impact. These results suggest a tangible synergistic relationship between homes within a residential development subjected to extreme winds. Although this synergistic relationship can exhibit a beneficial effect by reducing the wind loads experienced by any one singular building within a development, it can also produce a detrimental effect through the production of wind-borne debris. These results seem to agree with post-hurricane damage assessments that identify wind-borne debris impacts as a major contributor to building envelope damage. However, further studies should be carried out to develop fully the nature of this synergistic effect between homes within a typical residential development.

## References

Baker, C. J. (2007). "The debris flight equations." *J. Wind Eng. Ind. Aerodyn.*, **95**(5), 329–353.

Dao, T. N., van de Lindt, J. W., Prevatt, D. O., and Gupta, R. (2012). "Probabilistic procedure for wood-frame roof sheathing panel debris impact to windows in hurricanes." *Eng. Struct.*, **35**, 178–187.

Dunlap, J. A., and Widener, K. (1971). "Nuclear power plant tornado design considerations." *J. Power Div.*, **97**(2), 407–417.

FEMA. (2007). "Multi-hazard loss estimation methodology—Hurricane model." ⟨www.fema.gov/media-library-data/20130726-1820-25045-8522/hzmh2_1_hr_tm.pdf⟩ (Dec. 08, 2017).

Grayson, J. (2014). "Building envelope failure assessment of residential developments subjected to hurricane wind hazards." Ph.D. dissertation, Clemson Univ., Clemson, SC.

Grayson, J. M., and Pang, W. (2015). "Analytical damage quantification method for residential developments subjected to hurricane wind hazards." *Proc., 12th Int. Conf. on Applications of Statistics and Probability in Civil Engineering (ICASP12)*, T. Haukaas, ed., Univ. of British Columbia Library, Vancouver, Canada.

Grayson, M., Pang, W., and Schiff, S. (2012). "Three-dimensional probabilistic wind-borne debris trajectory model for building envelope impact risk assessment." *J. Wind Eng. Ind. Aerodyn.*, **102**, 22–35.

Grayson, J. M., Pang, W., and Schiff, S. (2013). "Building envelope failure assessment framework for residential communities subjected to hurricanes." *Eng. Struct.*, **51**, 245–258.

Gurley, K., et al. (2005). *Florida public hurricane loss projection model engineering team final report*, Florida International Univ., Miami, FL.

Haan, F. L., Sarkar, P. P., and Gallus, W. A. (2008). "Design construction and performance of a large tornado simulator for wind engineering applications." *Eng. Struct.*, **30**(4), 1146–1159.

Holmes, J. D. (2004). "Trajectories of spheres in strong winds with application to wind-borne debris." *J. Wind Eng. Ind. Aerodyn.*, **92**(1), 9–22.

Holmes, J. D. (2010). "Windborne debris and damage risk models: A review." *Wind Struct.*, **13**(2), 95–108.

Holmes, J. D., Baker, C. J., and Tamura, Y. (2006a). "Tachikawa number: A proposal." *J. Wind Eng. Ind. Aerodyn.*, **94**(1), 41–47.

Holmes, J. D., Letchford, C., and Lin, N. (2006b). "Investigations of plate-type windborne debris. II: Computed trajectories." *J. Wind Eng. Ind. Aerodyn.*, **94**(1), 21–39.

Kordi, B., Traczuk, G., and Kopp, G. A. (2010). "Effects of wind direction on the flight trajectories of roof sheathing panels under high winds." *Wind Struct.*, **13**(2), 145–167.

Kosiba, K. A., Trapp, R. J., and Wurman, J. (2008). "An analysis of the axisymmetric three-dimensional low-level wind field in a tornado using mobile radar observations." *Geophys. Res. Lett.*, **35**(5), L05805.

Lee, A. J. H. (1974). "A general study of tornado-generated missiles." *Nucl. Eng. Des.*, **30**(3), 418–433.

Lin, N., Holmes, J. D., and Letchford, C. W. (2007). "Trajectories of wind-borne debris in horizontal winds and applications to impact testing." *J. Struct. Eng.*, **133**(2), 274–282.

Lin, N., Letchford, C., and Holmes, J. D. (2006). "Investigation of plate-type windborne debris. Part I. Experiments in wind tunnel and full scale." *J. Wind Eng. Ind. Aerodyn.*, **94**(2), 51–76.

Lin, N., and Vanmarcke, E. (2008). "Windborne debris risk assessment." *Probab. Eng. Mech.*, **23**(4), 523–530.

Lin, N., and Vanmarcke, E. (2010). "Windborne debris risk analysis. I: Introduction and methodology." *Wind Struct.*, **13**(2), 191–206.

Lin, N., Vanmarcke, E., and Yau, S. C. (2010). "Windborne debris risk analysis. II. Application to structural vulnerability modeling." *Wind Struct.*, **13**(2), 207–220.

Maruyama, T. (2011). "Simulation of flying debris using a numerically generated tornado-like vortex." *J. Wind Eng. Ind. Aerodyn.*, **99**(4), 249–256.

McDonald, J. R., Mehta, K. C., and Minor, J. E. (1974). "Tornado-resistant design of nuclear power-plant structures." *Nucl. Safety*, **15**(4), 432–439.

Mishra, A. R., James, D. L., and Letchford, C. W. (2008). "Physical simulation of a single-celled tornado-like vortex. A: Flow field characterization." *J. Wind Eng. Ind. Aerodyn.*, **96**(8–9), 1243–1257.

Noda, M., and Nagao, F. (2010). "Simulation of 6DOF motion of 3D flying debris." *Proc., 5th Int. Symp. on Computational Wind Engineering (CWE2010),* Chapel Hill, NC. International Association for Wind Engineering (IAWE), Kanagawa, Japan.

Pinelli, J.-P., Pita, G., Gurley, K., Torkian, B., Hamid, S., and Subramanian, C. (2011). "Damage characterization: Application to Florida public hurricane loss model." *Nat. Hazards Rev.*, **12**(4), 190–195.

Redmann, G. H., Radbill, J. R., Marte, J. E., Degarabedian, P., and Fendell, F. E. (1976). "Wind field and trajectory models for tornado-propelled objects." *Tech. Rep. No. 1*, Electrical Power Research Institute, Palo Alto, CA.

Redmann, G. H., Radbill, J. R., Marte, J. E., Degarabedian, P., and Fendell, F. E. (1978). "Wind field and trajectory models for tornado-propelled objects." *Final Rep. 2*, Electrical Power Research Institute, Palo Alto, CA.

Richards, P. J., Williams, N., Laing, B., McCarty, M., and Pond, M. (2008). "Numerical calculation of the three-dimensional motion of wind-borne debris." *J. Wind Eng. Ind. Aerodyn.*, **96**(10–11), 2188–2202.

Simiu, E., and Cordes, M. (1976). "Tornado-borne missile speeds." NBSIR 76-1050, National Bureau of Standards, Gaithersburg, MD.

Sparks, P. R., and Huang, Z. (2001). "Gust factors and surface-to-gradient wind-speed ratios in tropical cyclones." *J. Wind Eng. Ind. Aerodyn.*, **89**(11–12), 1047–1058.

Tachikawa, M. (1983). "Trajectories of flat plates in uniform flow with application to wind-generated missiles." *J. Wind Eng. Ind. Aerodyn.*, **14**(1–3), 443–453.

Tachikawa, M. (1988). "A method for estimating the distribution range of trajectories of wind-borne missiles." *J. Wind Eng. Ind. Aerodyn.*, **29**(1–3), 175–184.

Twisdale, L. A., Dunn, W. L., and Davis, T. L. (1979). "Tornado missile transport analysis." *Nucl. Eng. Des.*, **51**(2), 295–308.

Twisdale, L. A., Vickery, P. J., and Steckley, A. C. (1996). "Analysis of hurricane windborne debris impact risk for residential structures." *Rep. 5303*, Applied Research Associates, Raleigh, NC.

Vickery, P. J., Lin, J., Skerlj, P. F., Twisdale, L. A., and Huang, K. (2006a). "HAZUS-MH hurricane model methodology I: Hurricane hazard, terrain, and wind load modeling." *Nat. Hazards Rev.*, **7**(2), 82–93.

Vickery, P. J., Skerlj, P. F., Lin, J., Twisdale, L. A., Young, M. A., and Lavelle, F. M. (2006b). "HAZUS-MH hurricane model methodology II: Damage and loss estimation." *Nat. Hazards Rev.*, **7**(2), 94–103.

Vickery, P. J., Wadhera, D., Twisdale, L. A., and Lavelle, F. M. (2009). "U.S. hurricane wind speed risk and uncertainty." *J. Struct. Eng.*, **135**(3), 301–320.

Visscher, B. T., and Kopp, G. A. (2007). "Trajectories of roof sheathing panels under high winds." *J. Wind Eng. Ind. Aerodyn.*, **95**(8), 697–713.

Wills, J. A. B., Lee, B. E., and Wyatt, T. A. (2002). "A model of wind-borne debris damage." *J. Wind Eng. Ind. Aerodyn.*, **90**(4–5), 555–565.

Yau, S. C., Lin, N., and Vanmarcke, E. (2011). "Hurricane damage and loss estimation using an integrated vulnerability model." *Nat. Hazards Rev.*, **12**(4), 184–189.

# CHAPTER 7

# Current Standards for Wind-Borne Debris Testing and Protection

Chris Letchford*

## 7.1 RATIONALE FOR DEBRIS IMPACT STANDARDS

From the earliest times, it was recognized that wind broke or picked up objects (tree limbs, sand, building components, etc.) and transported these objects over some distance, developing significant speeds prior to impact. With the development of cladding items that were susceptible to damage under impact, glazing in particular, and the knowledge that the impact occurred on windward faces, which could lead to internal pressurization of a building and consequent potential doubling of wind loads on the roof and other walls, clearly structures needed to be designed for full internal pressurization or vulnerable cladding needed to be protected. The almost infinite variety of debris items combined with the large range of wind speeds led to a wide range in impact momentum and energy, and so representative debris items were sought to create testing protocols to compare relative product performances under impact loading. The research described above has been used as the basis for developing the standard debris items (Chapter 2), standard impact energy and momentum (Chapters 4 and 5), and wind zones in which wind-borne debris poses a significant risk (Chapters 3 and 6).

It was acknowledged that quantifying the absolute risk of failure would be difficult to achieve but that relative failure risk would suffice given appropriately constructed impact tests. Observations of damage led to various debris classification schemes and the ability to create a representative test channeled testing protocols into simple missiles (steel balls, timber planks, or poles) that could be easily launched (cannons) at repeatable speeds. Early classification schemes (e.g., Bailey 1984, Lee 1974) divide the debris panoply into small, medium, and heavy-weight items, e.g., steel ball bearing, 2 × 4 lumber, and automobile. Lately, shape has become an important consideration due to debris aerodynamics and classification is

*Department of Civil and Environmental Engineering, Rensselaer Polytechnic Institute, Troy, NY, USA.

described as compact [three-dimensional (3D)], plate-like [two-dimensional (2D)], or rod [one-dimensional (1D)] (Wills et al. 2002). Again, the simplest model of 3D and 1D debris items are a steel ball bearing and 2 × 4 lumber. Plate-like debris items have not been pursued in impact tests because of the difficulty of launching such items.

## 7.2 ORIGINS OF WIND-BORNE DEBRIS TESTING STANDARDS

The initial efforts to quantify and design for debris impact involved high-risk facilities associated with the nuclear power industry. These were rightly very conservative. After substantial tornado damage to domestic construction and consequent loss of life, an industry began to develop around providing shelters, which necessitated impact resistance design criteria (Kiesling et al. 1974). Significant damage in hurricanes around the same time (Walker et al. 1975) also provoked the need to consider debris impact and envelope breach as design considerations for ordinary buildings.

### 7.2.1 Standard Review Plan (SRP) 3.5.1.4 (U.S. Nuclear Regulatory Commission 1975)

SRP 3.5.1.4 (U.S. Nuclear Regulatory Commission 1975) provided acceptance criteria for appropriate missiles and impact speeds for nuclear power plants. Seven different missiles ranging from a 4 kg (9 lb) steel rod to a 1.8 t (4,000 lb) automobile traveling at speeds ranging from 77 km/h (48 mi/h) (automobile) to 468 km/h (291 mi/h) (plank) are described. These were acknowledged as very conservative as appropriate for the nuclear power industry. After undertaking a more risk-consistent design and using actual field observations, these conservative missile criteria were amended to those presented in DOE STD 1020-94 (DOE 1994) and described in Kuilanoff et al. (1991).

### 7.2.2 DOE STD 1020-94 (DOE 1994)

DOE STD 1020-94 (DOE 1994) specifies design missile criteria for the range of extreme winds, from hurricanes, to thunderstorms, to tornadoes and was developed by a panel of experts. The basic missile is a 6.8 kg (15 lb), 2 × 4 lumber with speeds of 80 km/h (50 mi/h), (straight line winds), 161 km/h (100 mi/h), or 214 km/h (150 mi/h) depending on the performance category of the structure in a tornado. In addition, for tornadoes, a 34 kg (75 lb), 7.6 cm (3 in.) diameter steel pipe traveling at 80 km/h (50 mi/h) or 121 km/h (75 mi/h) is specified for tornadoes as well as the 1.4 t (3,000 lb) automobile traveling at 40 km/h (25 mi/h).

In a report for the Lawrence Livermore National Laboratory (LLNL), McDonald (1999) specifically developed tornado missile guidelines for the design of DOE facilities. Findings and recommendations were based on significant and extensive post-damage documentation studies from some 70 storms over a 20-year period. Missiles were classified as lightweight (lifted and transported by wind), medium-weight (accelerated horizontally while falling under gravity), and heavy (roll and tumble along the ground). Statistical analysis of the storms was undertaken to establish a probability distribution for each missile classification. The three categories were modeled by a 6.8 kg (15 lb) 2 × 4 lumber, a 34 kg (75 lb) 3-in.-diameter steel pipe, and a 1.4 t (3,000 lb) automobile, with the latter only evident in tornadoes. Missile trajectory simulations were developed but were probabilistically based. The simulations were apparently validated against Fujita-scale wind speed estimates from field observations of damage. The development of these standards is described in McDonald et al. (1974), Redmann et al. (1976), and Simiu and Cordes (1976).

### 7.2.3 Technical Record 440 (TR440) (Experimental Building Station 1978)

Following the devastation of the Australian city of Darwin by Tropical Cyclone Tracy in December 1974, a major review of design standards for cyclone regions in Australia was undertaken. One of the outcomes was the need to consider debris impact, cyclic loading, and subsequent fatigue failure of cladding elements. Immediately after the event the Darwin Area Building Manual (DABM 1975) recommended that to overcome the possibility of full internal pressurization by debris impact the building envelop must resist a "... 4k g (9 lb) piece of 50 mm by 100 mm (2 × 4 lumber) travelling at 72 km/h (45 mph) from causing an opening failure." Subsequently, a workshop held in July 1977 (Experimental Building Station 1978) with representation from industry, design consultants, researchers, and governmental authorities reached consensus on loads and testing procedures for the performance of building materials in cyclone-prone areas in Australia.

Of particular interest here is the impact testing developed for tropical cyclone-prone areas to assess the impact resistance of residential buildings or parts thereof:

> A 4 kg (9 lb) mass having a timber impacting head of nominal 50 mm by 100 mm (2 × 4 lumber) dimensions shall be used striking with an impact velocity of 54 km/h (34 mph). Where testing limitations make it desirable to use an increased mass at lower velocities, this may be done provided the missile has the same energy at impact, the velocity is not decreased by more than 25% and the specified impacting head is used (Experimental Building Station 1978).

The workshop report notes that "the prescribed criteria should not be regarded as providing guaranteed resistance against debris in a cyclone but rather as providing a performance standard for rating purposes which it is believed will provide reasonable protection against penetration." (Section A.5 of TR440 details the test assembly, loading, testing procedure, evaluation, and reporting of test results.)

### 7.2.4 Standards Association of Australia (SAA) AS-1170.2 (Standards Australia 1989)

In the 1989 edition of this code (Standards Australia 1989), buildings in cyclonic regions (northern coastlines up to 100 km from the coast) had to be designed for dominant openings (full internal pressurization) unless windows were capable of resisting a 4kg (9 lb) piece of timber of 100 mm × 50 mm cross section, striking them at any angle at a speed of 54 km/h (34 mi/h). This requirement has its origin in the TR440 Report (Experimental Building Station 1978). In the 2011 edition of this standard (Standards Australia 2011) loading from wind-borne debris is specified as having impact equivalent to (1) a 4 kg (9 lb) timber member 100 mm × 50 mm (3.9 × 2 in.) in a cross section impacting end on at $0.4V_R$ for horizontal trajectories and $0.1V_R$ for vertical trajectories and (2) a 8 mm (0.3 in.) diameter spherical steel ball of approximately 2 g (0.071 oz) impacting at $0.4V_R$ for horizontal trajectories and $0.3V_R$ for vertical trajectories. Here, $V_R$ is the regional wind speed, which is the nominal 3-s gust at 10 m (33 ft) in flat open terrain, with the appropriate return period chosen. These impact design conditions are applicable for design of shelters and for protection of openings in cyclonic regions to prevent full internal pressurization. The change to a variable impact speed reflects recent studies (e.g., Letchford et al. 2013) that developed actual asymptotic flight speeds for debris items.

### 7.2.5 South Florida Building Code (SFBC)

After Hurricane Andrew in 1992, surveys showed that damage caused by missile debris impact and internal pressurization was extensive. Missile debris was listed as one of the principal reasons for poor building performance in Hurricane Andrew after roofing failures. Several organizations issued reports with recommendations regarding the important role of missile debris, windward wall openings, and internal pressure on the performance of buildings during the hurricane, as shown in Table 7.1. These events finally led to the adoption of a debris impact standard in the SFBC in Broward County (February 10, 1994) (1994a), in Dade County (December 14, 1993) (1994b), and Palm Beach County (February 20, 1996). It was later incorporated into the Standard Building Code.

Since September 1, 1994, the Dade County and Broward County editions of the SFBC require windows, doors, and wall coverings to be designed for wind-borne missile debris impacts. Under these codes, all wall elements must pass county approved flying missile impact and wind loading tests. These tests are designed to replicate hurricane conditions and to test the ability of windows, shutters, and all openings to maintain the integrity of the building envelope.

The test protocols in Dade and Broward Counties defined a large missile as a 4kg (9 lb) 2 × 4 lumber impacting at a speed of 55 km/h (34 mi/h) for building elevations below 9.1 m (30 ft). For a small missile, there are differences between the Dade and Broward County editions in the missile type and speed of impact. The Dade County edition specifies 10 pieces of 2 g (0.071 oz) roof gravel impacting simultaneously at a speed of 88 km/h (55 mi/h) while the Broward County edition specifies 10 pieces of 2 g (0.071 oz) steel ball impacting successively at 143 km/h

*Table 7.1. Direct Extracts from Major Reports on Hurricane Andrew [Behr and Minor (1994)]*

| *Investigative authority (reference)* | *Observations* | *Recommendations* |
|---|---|---|
| Building Code Evaluation Task Force (Dade County 1992a) | Penetration of building envelopes unprotected and vulnerable to wind-borne debris. | Adopt mandatory protection of exterior building envelopes by requiring that windows be specifically designed for impact loads from wind-borne debris. |
| Dade County Grand Jury (Dade County 1992b) | The use of materials, particularly for the exterior integrity of a building, needs to be reviewed. | The code must include the mandatory use of storm shutters or a tested acceptable alternative. |
| Florida Department of Community Affairs (1997) | Flying debris contributed to a significant portion of (window) damage. | Windows shall resist missile impact or to be protected from missile impact. |
| Federal Insurance Administration (FEMA 1992) | The breaching of the building envelope by failures of (windows) due to debris impact was a significant factor in the damage. | Window design and protection standards should be developed. Consider using shutters and precut plywood for protection of existing buildings. Garage door assemblies should be strengthened. |
| Wind Engineering Research Council (1992) | Many high-rise buildings sustained breakage of windows due to primarily wind-borne debris. | Provisions should be adopted to account for increases in internal pressure due to breaching of building envelope by wind-borne debris or requirements should be added to protect windows. |

*Table 7.2. Test Protocols for Small and Large Missiles in SFBC*

| | *Large missile* | *Small missile* | |
|---|---|---|---|
| *Parameter* | *Dade and Broward Counties* | *Dade County* | *Broward County* |
| Building elevation | <9 m (30 ft) | >9 m (30 ft) | >9 m (30 ft) |
| Number of pieces and impact type | 1 | 10 (simultaneously) | 10 (successively) |
| Missile type | 2 × 4 lumber | Roof gravel | Steel ball |
| Missile weight | 4 kg (9 lb) | 2 g (0.071 oz) | 2 g (0.071 oz) |
| Impact speed | 15 m/s (50 ft/s) | 24 m/s (80 ft/s) | 40 m/s (130 ft/s) |

(89 mi/h). The small and large missile impact tests are followed by a 9,000-cycle pressure spectrum (cyclic pressure loading test). The test protocols have been summarized in Table 7.2.

### 7.2.6 Standard Building Code (SBC)

The Southern Building Code Congress International (SBCCI), representing 25 states, made recommendations for stronger shutters and windows that can meet a new missile impact test. The SBCCI proposal recommended that buildings either be designed for full internal pressure, or have external protection devices (shutters) for windows, or have impact resistant glazing. The proposed changes to the SBC require that glazed openings and doors in buildings have external impact protective systems or exterior impact resistant glazed systems. Shutters and impact resistant glazing would be qualified by the SBCCI (1999) test standard for determining impact resistance from missile debris (SSTD 12-99).

In 1983, after Hurricane Alicia caused major damage to architectural glazing systems in Houston, a requirement that windows in hurricane-prone regions be designed for missile debris impact was proposed as a change to the SBC in the United States. This proposal was not adopted.

### 7.2.7 International Building Code and International Residential Code

In 2003, the Building Officials Code Administrators International, SBCCI, and International Conference of Building Officials combined to create the International Code Council (ICC), which publishes a single family of codes including the International Building Code (IBC). Section 1609 of the IBC contains wind-load provisions and specifies the following requirements of ASTM E1886-13 (ASTM 2013) and ASTM E1996-17 (ASTM 2017) for glazing protection in wind-borne debris regions. It also specifies using a large missile impact test in accordance with ASTM E1996, if glazed openings are located less than 9.14 m (30 ft) from the ground. It specifies using the small missile impact test in accordance with ASTM E1886, if glazed openings are located higher than 9.14 m (30 ft.). Section R609.6 of the 2015 International Residential Code (ICC 2015) specifies use of ASTM E1886, ASTM E1996, and AAMA 506 for testing and labeling of exterior windows, doors, or other fenestration products if buildings are located in wind-borne debris regions.

### 7.2.8 ASCE 7 (ASCE 2010)

This standard, ASCE 7 (ASCE 2010), specifies that glazing in certain regions of the United States needs to be protected against debris impact. These regions are where the basic wind speed exceeds 225 km/h (140 mi/h) or is greater than 209 km/h (130 mi/h) and within 1.6 km (1 mi) of the coastal high waterline. Typically, this confines such protection to severe hurricane regions. Any protection for such glazing must conform to ASTM E-1886 and ASTM E-1996 testing requirements.

### 7.2.9 ASTM E-1886-13a (ASTM 2013) and ASTM E-1996-17 (ASTM 2017)

An effort to standardize debris impact testing was undertaken in the United States, with ASTM revising standards E-1886 and E-1996 in 2013 and 2017 respectively. These standards cover specification and test methods based on performance requirements of the IRC and application requirements of the IBC aimed at assessing protection of exterior windows, curtain walls, doors, and impact protection systems that are also subject to cyclic loading as might be expected during the passage of a hurricane. They specify a small missile (2 g) steel ball traveling at 0.4 to 0.85 times the basic wind speed [3-s gust at 10 m (33 ft) in flat open terrain] and a large missile (2 × 4 lumber) with weights ranging from 0.9 kg (2 lb) to 6.8 kg (15 lb) and speeds ranging from 0.1 to 0.55 of the basic wind speed for the location in question.

## 7.3 COMPARISON OF DEBRIS IMPACT TESTING STANDARDS

Multiple associations have their own criteria for wind-borne debris testing such as the ASCE, FEMA, DOE, ASTM International, ICC, and SAA. All, except SAA, directly or indirectly specify ASTM standards to perform wind-borne debris impact tests. In the United States, ASTM developed its early wind-borne debris tests from some of Australia's early testing methods. The U.S. DOE, however, needed higher impact tests for stricter building requirements for critical buildings categorized to need high resistance to wind-borne debris damage. According to FEMA-361 (Chapter B8, FEMA 2015), there has been considerable support due to legacy codes and standards to use a 4 kg (9 lb), 2 × 4 lumber as the missile of choice for hurricanes, which dates back to the 1976 Darwin Area Building Manual (DABM 1975) in Australia, created in response to Cyclone Tracy. FEMA also states that in 1999 FEMA and Texas Tech University selected a 6.8 kg (15 lb), 2 × 4 lumber to be the representative tornado missile in the United States. The DOE also adopted the 6.8 kg (15 lb) 2 × 4 lumber as its missile of choice. Depending on the standard being followed for tornadoes or hurricanes the missile may be different, with the 4 kg (9 lb) missile typically used for hurricanes and cyclones (see Table 7.3 for a summary of the different testing requirements).

*Table 7.3 Comparison of Debris Impact Test Requirements for Tornadoes and Hurricanes*

| *Guidance, code of standard criteria for the design missile* | *Horizontal debris impact test speed km/h (mi/h)* | *Large missile Specimen kg (lb)* | *Momentum at impact kg·m/s (lb·s)* | *Energy at impact joule (ft·lb)* |
|---|---|---|---|---|
| | Tornado safe room missile testing requirements | | | |
| DOE-STD-1020-2002 (DOE 2002) | 40 (25) | 1,360 (3,000) auto | 14,400 (3,240) | 91,800 (67,710) |
| | 80 (50) | 34 (75) pipe | 761 (171) | 8,500 (6,270) |
| | 121 (75) | 34 (75) pipe | 1,140 (257) | 19,100 (14,100) |
| | 241 (150) (max.) | 6.8 (15) 2 × 4 | 458 (103) | 15,300 (11,290) |
| | 161 (100) (min.) | 6.8 (15) 2 × 4 | 302 (68) | 6,800 (5,020) |
| FEMA-320 (FEMA 2014)/FEMA-361 (FEMA 2015) | 161 (100) (max.) | 6.8 (15) 2 × 4 | 302 (68) | 6,800 (5,020) |
| | 129 (80) (min.) | 6.8 (15) 2 × 4 | 245 (55) | 4,350 (3,210) |
| ICC-500 storm shelter standard (ICC 2013) | 161 (100) (max.) | 6.8 (15) 2 × 4 | 302 (68) | 6,800 (5,020) |
| | 129 (80) (min.) | 6.8 (15) 2 × 4 | 245 (55) | 4,350 (3,210) |
| IRC/IBC (ICC 2015a, b), ASCE 7-10, Florida and North Carolina State Building Codes, ASTM E1886 (ASTM 2005b)/ASTM E1996 (ASTM 2005a) | N/A | N/A | N/A | N/A |
| | Hurricane safe room missile testing requirements | | | |
| DOE-STD-1020-2002 (DOE 2002) | 80 (50) | 6.8 (15) 2 × 4 | 151 (34) | 1,700 (1,254) |
| FEMA-320 (FEMA 2014)/FEMA-361 (FEMA 2015) | 206 (128) (max.) | 4 (9) 2 × 4 | 236 (53) | 6,690 (4,930) |
| | 129 (80) (min.) | 4 (9) 2 × 4 | 147 (33) | 2,610 (1,930) |

| | | | | |
|---|---|---|---|---|
| ICC-500 storm shelter standard (ICC 2013) | 102 (max.) | 4 (9) 2 × 4 | 187 (42) | 4,250 (3,130) |
| | 64 (min.) | 4 (9) 2 × 4 | 116 (26) | 1,670 (1,230) |
| Florida State Emergency Shelter Program Criteria and EOC Design Criteria | 80 (50) (EOC recommended) | 6.8 (15) 2 × 4 | 151 (34) | 1,700 (1,250) |
| | | 4 (9) 2 × 4 | 102 (23) | 1,235 (911) |
| | 88 (55) (EHPA recommended) | 4 (9) 2 × 4 | 62 (14) | 472 (348) |
| | 55 (34) (EHPA min.) | | | |
| IBC/IRC 2012, ASCE 7-10 Florida and North Carolina State Building Codes, ASTM E1886 (ASTM 2005b)/ASTM E1996 (ASTM 2005a) | 88 (55) | 4 (9) 2 × 4 | 102 (23) | 1,234 (910) |
| | 55 (34) | 4 (9) 2 × 4 | 62 (14) | 472 (348) |
| ICC 600-14 | 55 (34) | 4 (9) 2 × 4 | 62 (14) | 472 (348) |

*Note*: max. = maximum; min. = minimum; N/A = not available.

## 7.4 SUMMARY

Wind-borne missile debris, the major overlooked aspect of the hurricane and tornado environment, has found its way into building codes, test standards, and design practice. Several Florida counties, including Dade, Broward, and Palm Beach Counties have adopted the debris impact standard in the SFBC. Wind-borne debris impact protection is addressed in the IBC, IRC, and ICC 600.

Wind-borne debris impacting a building during a high wind storm or hurricane can puncture the envelope, and punctures in the building envelope can lead to damage to interior finishes and contents of the structure as a result of wind-driven water infiltration through the envelope. Such events are generally associated with hurricanes and coastal storms where missile impacts can result in extensive and costly damage to the interior of the structure. Therefore, in wind-borne debris regions, door and window assemblies must be specified to resist test missile loads specified in ASTM E1996-14a (ASTM 2014) and garage doors should conform to the requirements of ANSI/DASMA 115 (2012).

Although estimates can be made based on past observations and learning, there is still much scope for research on debris impact standards and quality building envelopes free of debris impact failure. Building codes should incorporate such standards to avoid fatalities and cases of building envelope breach as a result of missile debris impact.

### *References*

AAMA (American Architectural Manufacturers Association). "Voluntary specifications for impact and cycle testing of fenestration products." *AAMA 506*, Schaumburg, IL.

ASCE. (2010). "Minimum design loads for buildings and other structures." *ASCE/SEI 7-10*, Reston, VA.

ASTM. (2013). "Standard test method for performance of exterior windows, curtain walls, doors, and impact protective systems impacted by missile(s) and exposed to cyclic pressure differentials." *E1886-13a*, West Conshohocken, PA.

ASTM. (2017). "Standard specification for performance of exterior windows, glazed curtain walls, doors, and storm shutters impacted by windborne debris in hurricanes." *E1996-17*, West Conshohocken, PA.

Bailey, J. R. (1984). "Wall barrier resistance to the impact of tornado missiles." Master's thesis, Texas Tech Univ., Lubbock, TX.

Behr, R. A., and Minor, J. E. (1994). "A survey of glazing system behavior in multi-story buildings during Hurricane Andrew." *Struct. Des. Tall Special Build.*, **3**(3), 143–161.

Dade County. (1992a). "Building code evaluation task force final report." Metro Dade County Building Evaluation Task Force, Miami, FL.

Dade County. (1992b). "Final report of the Dade County grand jury." Miami, FL.

DOE (Department of Energy). (1994). "Natural phenomena hazards design and evaluation criteria for Department of Energy Facilities." ⟨https://nnsa.energy.gov/sites/default/files/nnsa/inlinefiles/doe%20standard%201020-2002.pdf⟩ (Nov. 2, 2017).

DOE. (2002). "Standard for natural phenomena hazards design and evaluation criteria for Department of Energy facilities." ⟨https://nnsa.energy.gov/sites/default/files/nnsa/inlinefiles/doe%20standard%201020-2002.pdf⟩ (Oct. 23, 2017).

Door and Access Systems Manufacturers Association. (2012). "Standard method for testing sectional garage doors and rolling doors: determination of structural performance under uniform static air pressure difference." ⟨http://www.dasma.com/dasma-pages/D-AS-standards.asp⟩ (Nov. 2, 2017).

Experimental Building Station. (1978). "Guidelines for the testing and evaluation of products for cyclone prone areas." *Tech. Rec. 440*, Dept. of Housing and Construction, Chatswood, Australia.

Federal Insurance and Mitigation Administration. (1992). *Building performance: Hurricane Andrew in Florida*, FEMA, Washington, DC.

FEMA. (2014). "Taking shelter from the storm: Building a safe room for your home or small business." ⟨https://www.fema.gov/media-library-data/1418837471752-920f09bb8187ee15436712a3e82ce709/FEMA_P-320_2014_508.pdf⟩ (Nov. 02, 2017).

FEMA. (2015). "Safe rooms for tornadoes and hurricanes: Guidance for community and residential safe rooms." ⟨https://www.fema.gov/media-library-data/1467990808182-0272256cba8a35a4e8c35eeff53dd547/fema_p361_July2016_508.pdf⟩ (Nov. 02, 2017).

Florida Division of Emergency Management. (2016). "2016 Statewide emergency shelter plan." ⟨http://www.floridadisaster.org/Response/engineers/SESPlans/2016SESPlan/index.html⟩ (Nov. 7, 2017).

ICC (International Code Council). (2013). "ICC draft standard on design, construction, and performance of storm shelters." ⟨https://www.iccsafe.org/cs/standards/IS-STM/Documents/2013Dev/ICC500-2013_PC1.pdf⟩ (Nov. 11, 2017).

ICC. (2014). "Standard for residential construction in high-wind regions." *ICC 600-2014*, Washington, DC.

ICC. (2015a). "International residential code." ⟨https://codes.iccsafe.org/public/document/toc/542/⟩ (Nov. 11, 2017).

ICC. (2015b). "International building code." ⟨https://codes.iccsafe.org/public/document/toc/553/⟩ (Nov. 11, 2017).

Kiesling, E. W., McDonald, J. R., Mehta, K. C., and Minor, J. E. (1973). "Residential buildings engineered to resist tornadoes." *J. Struct. Div.*, **99**(4), 1314–1316.

Lee, A. J. H. (1974). "A general study of tornado-generated missiles." *Nucl. Eng. Des.*, **30**(3), 418–433.

Letchford, C., Lin, N., and Holmes, J. (2013). "Windborne debris in horizontal winds and applications to impact testing." Chapter 8, *Advanced structural wind engineering*, Y. Tamura and A. Kareem, eds., Springer, Tokyo, 197–215.

McDonald, J. R. (1999). "Rationale for wind-borne missile criteria for DOE facilities." *UCRL-CR-135687*, Lawrence Livermore National Laboratory, Livermore, CA.

McDonald, J. R., Mehta, K. C., and Minor, J. E. (1974). "Tornado-resistant design of nuclear power-plant structures." *Nucl. Safety*, **15**(4), 432–439.

Minor, J. E. (1994). "Windborne debris and the building envelope." *J. Wind Eng. Ind. Aerodyn.*, **53**(1), 207–227.

Redmann, G. H., Radbill, J. R., Marte, J. E., Degarabedian, P., and Fendell, F. E. (1976). "Wind field and trajectory models for tornado-propelled objects." *Tech. Rep. No. 1*, Electrical Power Research Institute, Palo Alto, CA.

Simiu, E., and Cordes, M. (1976). "Tornado-borne missile speeds." *NBSIR 76-1050*, National Bureau of Standards, Gaithersburg, MD.

SFBC (South Florida Building Code). (1993). "Metropolitan Dade County edition of the South Florida Building Code." Metropolitan Dade County, Miami, FL.

SFBC. (1994a). "Broward County edition of the South Florida Building Code." Broward County, Ft. Lauderdale, FL.

SFBC. (1994b). "Metropolitan Dade County edition of South Florida Building Code." Metropolitan Dade County, Miami, FL.

SBCCI (Southern Building Code Congress International). (1999). "Standard for determining impact resistance from wind borne debris." *SBCCI SSTD 12-99*, Birmingham, AL.

Standards Australia. (1989). "SAA loading code. 2: Wind loads." *AS 1170.2-1989*, North Sydney, Australia.

Standards Australia. (2011). "SAA loading code. 2: Wind loads." *AS 1170.2*, North Sydney, Australia.

University of Florida School of Building Construction, Department of Community Affairs, Division of Emergency Management. (1997). "Hurricane evacuation shelter selection guidelines student manual." ⟨http://www.floridadisaster.org/Response/engineers/HES.htm⟩ (Dec. 08, 2017).

U.S. Nuclear Regulatory Commission. (1975). "Standard review plan for the review of safety analysis reports for nuclear power plants: LWR edition—Design of structures, components, equipment, and systems." ⟨https://www.nrc.gov/reading-rm/doc-collections/nuregs/staff/sr0800/ch3/⟩ (Nov. 2, 2017).

Walker, G. R., Minor, J. E., and Marshall, R. D. (1975). "Darwin cyclone—Valuable lesson in structural design." *Civil Eng.*, **45**(12), 82–86.

WERC (Wind Engineering Research Council). (1992). "Preliminary observations of WERC post-disaster team." College Station, TX.

Wills, J. A. B., Lee, B. E., and Wyatt, T. A. (2002). "A model of wind-borne debris damage." *J. Wind Eng. Ind. Aerodyn.*, **90**(4–5), 555–565.

# Index